WEEDS

Reaktion's Botanical series is the first of its kind, integrating horticultural and botanical writing with a broader account of the cultural and social impact of trees, plants and flowers.

Published
Apple Marcia Reiss
Ash Edward Parker
Bamboo Susanne Lucas
Berries Victoria Dickenson
Birch Anna Lewington
Cactus Dan Torre
Cannabis Chris Duvall
Carnation Twigs Way
Carnivorous Plants Dan Torre
Cherry Constance L. Kirker and Mary Newman
Chrysanthemum Twigs Way
Geranium Kasia Boddy
Grasses Stephen A. Harris
House Plants Mike Maunder
Lily Marcia Reiss
Moss and Lichen Elizabeth Lawson
Mulberry Peter Coles
Oak Peter Young
Orchid Dan Torre
Palm Fred Gray
Pine Laura Mason
Poppy Andrew Lack
Primrose Elizabeth Lawson
Rhododendron Richard Milne
Rose Catherine Horwood
Rowan Oliver Southall
Snowdrop Gail Harland
Sunflowers Stephen A. Harris
Tulip Celia Fisher
Weeds Nina Edwards
Willow Alison Syme
Yew Fred Hageneder

WEEDS

Nina Edwards

REAKTION BOOKS

Published by
REAKTION BOOKS LTD
Unit 32, Waterside
44–48 Wharf Road
London N1 7UX, UK

www.reaktionbooks.co.uk

First published 2015
First published in paperback 2024

Printed and bound in India by Replika Press Pvt. Ltd

A catalogue record for this book is available from the British Library

ISBN 978 1 78914 958 6

Contents

Weeds grow everywhere.

Introduction

Weeds are everywhere. We may try to bend nature to our will, but on the peripheries of herbicide-drenched fields, squeezed between paving slabs and seeded in the crevices of city walls, these often disregarded plants grow and prosper. They are the tough underclass of the plant world, bullying their way into drainage systems, poisoning meadowland, stealing essential nutrients and taking a stranglehold on plants we deem legitimate. Yet in those far-flung places where wilderness remains we often do not bother to consider weeds at all. And they require our attention. Weeds exist only in relation to ourselves.

'Weed' is a word that conceals inherent contradictions. In German it is something less than other plants, the un-herb, the *Unkraut*, echoing the Nazi term for a Jew, *Unmensch*, an un-person. In French and Spanish it is a bad herb, a herb gone wrong, *mauvaise herbe*, *mala hierba*. The modern Italian *erbaccia* is also pejorative, suggesting something ugly and useless. In Latin *viriditas* is also less derogatory – simply a green thing. The English word 'weed' is derived from the Old Saxon *wiod*, related to a fern, a wild and prolific plant. Though the derivation of 'weeds' as a word for clothing is otherwise, from the Old Saxon *wad*, the term nonetheless suggests a sense of plentiful fabric, as in women's clothing, or dark, dishevelled, wild disregard, as in widows' mourning weeds. *Waedless* was a term for naked and *waed-bréc* for breeches. Edmund Spenser, in his *The Faerie Queene*, has 'A goodly Ladie clad in hunters weed' (Book 2, canto III.xxi).

The common names for weeds are robust, no-nonsense and plainly descriptive: pigweed, goutweed, snout, nosebleed. Turn to the Glossary and you will see that English terms for weed species are often Anglo-Saxon, and there is no beating about the bush with piss-a-bed, bridewort, fleabane or madwoman's milk. The Latin double-barrelled system has gravitas but lacks a certain attack. In its everyday name the function of or dread surrounding a particular plant is often laid bare.

For followers of Zen, weeds are treasures; for Christians they may act as a reminder of almighty design. In an introduction to Taoist philosophy, removing weeds to allow crops to flourish signifies the need to cleanse the mind, for 'unless they are removed, concentration and wisdom will not grow'.[1] Tolstoy drew on the idea of weeding as a metaphor for sin, advising against merely mowing down weeds in a meadow, for like evil, weeds had to be taken out by the very roots.[2]

When the novelist Deborah Moggach's mother began to pull up garden plants and leave the weeds behind, it was a sign that she was beginning to suffer from dementia.[3] Weeds can be a gauge of human behaviour, and in Shakespeare they suggest personal disorder, particularly in the sonnets, and political upheaval when weeds such as 'darnel, hemlock, and rank fumitory' are allowed to flourish in *Henry V* (Act V, Scene 2). If weeds are wild plants, plants out of our control, then those that inhabited the earth before our coming might all be so called.

Weeds, then, are despised, voracious plant life if found in the garden border, but in the wild, beyond our perception, when we cannot see or do not care to know what damage they may be doing, it is a matter of live and let live. The perceptive gardener may think twice before rooting up all weeds in their path, noticing what they reveal about soil fertility, whether it is acid or alkaline, and perhaps fearing that patch where nothing ever grows. The ecologist may reckon that the balance of plant life in a particular habitat has been threatened by the success of one species over others more worthy of existence, particularly if it is new to an area, proscribed as an alien invader thriving at the expense of other so-called indigenous plants

Camille Pissarro, *Woman Working in a Garden*, c. 1900, watercolour over black chalk.

Satan sowing darnel. Print by Pieter Jalhea Furnius after Gerard Groenning, 1585.

that have more right to exist there. And for a plant to be alien, it has to have been introduced by human activity, either intentionally or unintentionally, and so one might argue that weeds are our fault; yet we blame the weeds.

The idea of weeds challenges a romantic notion of there being a natural balance in vegetable and animal society, where plants live together harmoniously in a mutually beneficial set of relationships. If weeds create a persistent problem, they are easily thought of as criminal recidivists, carrying all the fear and loathing of their human counterparts, threatening what should be gentle, cooperative nature. Cooperative with our interests, that is. They crop up unexpectedly, and proliferate like sexually promiscuous immigrants, vying for available living space and greedily taking more than their share of available sun, rain, shelter and food. Many weeds reproduce asexually, of course, but the point here is that their powers of reproduction can seem a parallel with licentious human behaviour.

Weeds may be propagated with or without our connivance. Invasive weed plants can cause problems for wildlife, as with the Indian rhinoceros, Asian elephant and Bengal tiger in Assam, whose decline is linked to several weeds species.[4] Take camfhur grass as an example, a tropical shrub of the sunflower family whose rapid growth and easy reproduction by seed and root renders animal forage toxic. Each plant can produce 80,000 to 90,000 seeds. The plant is said to have first escaped from North America in the nineteenth century, not only to India from botanical gardens in Pakistan but also to African rainforests via its accidental addition to forestry seed. Although it was introduced to control cogon grass in pastureland in the Ivory Coast, in Sri Lanka it has infested coconut plantations and, like so many plants that remain relatively innocuous in more temperate regions, it has become a virulent weed across Australia.

Weeds is about the idea of weeds as much as the plants themselves. It is a topsy-turvy concept: many of our garden plants have the same biological characteristics as weeds. It shifts and turns about, depending entirely – or at least partly – upon context: 'They are weeds because they grow where they should not be. The gorgeous scarlet poppy is a weed amid the corn. If roses overgrew the wheat, we should dub them weeds, and root them out.'[5] Climatic and land-management changes and even the whims of gardening and wider cultural fashions alter what is considered a weed. Millet is an important crop in Canada, for example, but in the last 30 years it has also become a hostile agricultural weed in both Canada and America. Ryegrasses planted as forage or to avoid soil erosion, as in western Australia, where farmers alternate between wheat and sheep pasture, have become so well established that they have begun to compete with nitrogen-fixing clover and therefore to decrease subsequent wheat yields.[6] A study of Chilean rhubarb in Ireland, where its huge leaves tend to shade other plants, stopping them from germinating and growing, concluded that weed scientists know surprisingly little about the relationship between plant traits, habitat and weather.[7]

Weeds confront both our sense of a purposeful, designed world, and our confidence in one in which we should be the major players, as of right. Plants that are considered weeds are many and various, from minute proliferating unicellular algae dating back billions of years to mighty hybrid trees. But by looking at specific examples I hope to show how plants come to gain, and lose, and sometimes regain that label, and maybe even earn it permanently. Their history is played out, from our earliest attempts to manage nature to the successes of the Industrial Revolution and more lately to the problems resulting from intensive farming methods and its 'green' counter-reaction.

Anyone who has noticed a plume of buddleia in full purple glory reaching up out of a railway siding, or the appearance through spring of the yellow star flowers of coltsfoot, followed by celandine, cowslip and lady's smock, stitchwort decorating a ditch in May, water crowsfoot beginning to uncurl beside running water in June, water betony and meadowsweet, punctured by twisted spires of loosestrife with dragonflies hovering about – even a tiny yellow head of groundsel growing out of a broken gutter – will know that plants that can be seen as weeds are not necessarily unlovely. Yet there is a tendency to view what on occasion challenges our interests as both ugly and even morally deviant: 'not valued for use or beauty, growing wild and rank, and regarded as cumbering the ground or hindering the growth of superior vegetation'.[8]

Buddleia davidii, for example, was introduced into northern Europe from China in the 1890s as a prized exotic shrub, but when it ran wild on waste ground it was soon demoted to the rank of common weed. When it seeded and grew where it was not wanted, the beauty of its flowers was forgotten and its easy-virtue regenerative abilities became something to despise. We fear those plants that suggest an untidy world, urban decay, down-and-outs, falling property prices and all the grubby dissolution of economic struggle. In American towns weeds are 'happenstance plants', their presence suggesting infection and disease, drug dealers and illegal dumping, the heat-trapping properties of built-up areas producing more robust strains, infesting the

Dandelions thrive on a pavement.

disused buildings left behind by Hurricane Katrina or the Californian housing bust.[9]

How do we pick on some plants more than others? The native British common bluebell is less invasive than the Spanish bluebell because the woodland conditions in which it flourishes are far more limited; the Spanish is more adaptable and thus despised as cocky, pushy and foreign. The lesser celandine is rarely a threat to other plants because of its relatively short growing period, yet it can produce a shallow carpet of its tuber roots that get in the weedy way of other plants in the garden. While it is regarded as a harmless wild plant in hedgerows and woodland in the British Isles, it has become a menace in North America. Groundsel is welcome green food for caged birds, folklore suggests that the bitter perfume of its roots is a cure for

Buddleia taking root in a wall.

Sudama sits with his wife who urges him to seek Krishna's help. Outside their hovel datura weeds spring from the green courtyard, a symbol of poverty. Garhwal, India, *c.* 1775–90, artist unknown, painted in opaque watercolour on paper.

headaches and it has long been used as a diuretic, but the allotment holder applies the hoe to a plant that harbours leaf rust and a fungus that causes black root rot in peas. The alkaloids it contains remain active in silage, so that farmers who do not manage to keep it out of the food chain risk their stock suffering liver damage. A prejudice against groundsel might lie in its name, from the Anglo-Saxon *grounde-swelge*, ground-swallower, referring to its easy spreading habit, profligate and unstoppable. On the other hand, the origin of *grounde-swelge* is more usually given as 'pus-swallower', after its use in poultices.

A range of recent literature confounds a too-easy dismissal of weeds' usefulness. They have long fulfilled practical roles in many cultures, in particular as foraged food. Fat hen, for example, was used as a vitamin C-rich vegetable from Neolithic times right up until the sixteenth century, when even its seeds were a source of flour. Weeds can act as fertilizer; as an essential part of biodiversity; and as medicine, cosmetic, aphrodisiac or poison; for clothing, shelter and fuel; and so on. Some weeds attract wildlife to pollinate crops, destroy

An English bluebell, watercolour, 1903.

harmful insects and invite game. Artists are drawn to weeds as symbols of the irrepressible wild, as *rus in urbe*, the country in the town.

Search the Internet for weeds and you will find yourself reading about marijuana, or 'the weed', an affectionate term adopted in the 1960s when, perhaps, along with radical ideas about racial and sexual equality, there came a revaluation of a concept like weed and its former denigration. Cannabis, as the plant is also known, is indigenous to central and southern Asia. In dried form it is usually smoked and its resin can be made into cakes, which can be consumed, used for a tea or vaporized. In its most potent form it is called skunk. The psychoactive drug induces euphoria and relaxation, and on occasion intense paranoia. Its earliest recorded use is as far back as the third millennium BCE, having been found beside a mummified shaman in northwest Xinjiang, China, to see him on his way to the afterlife. This most popular of recreational drugs, sold as a harmless sweet in the West in the latter part of the nineteenth century as 'Arabian Gunje of Enchantment', has been widely legislated against since the early twentieth century, though lately there has been some softening of this attitude. Uruguay, for example, has legalized its production, marketing and consumption; in America, not only do 23 states allow its medical use, but in Colorado and Washington it is legal for recreational purposes. The argument continues as to whether or not it is a relatively harmless pleasure or a pernicious gateway to more serious drug use.

In classical literature the goddess Ceres created the poppy in order to eat its seeds and so forget her grief for the loss of her daughter, Proserpina. The opium poppy, *Papaver somniferum*, or sleep-inducing poppy, is the source of morphine and its derivative, heroin. The word heroin derives from the idea of the classical Greek hero, suggesting that it makes one feel extraordinarily powerful and courageous, or godlike. The opium or morphine poppy was first cultivated in Mesopotamia in 3400 BCE. The Sumerians knew it as *hul gil*, the joyful herb. It once grew on only a relatively narrow stretch of land through southern Asia, to Pakistan and Laos, but today much of it is produced in South America. Just as it is still grown for the beauty of its flowers in many gardens, its seed is traditionally used to top bread. Recently a Swiss traveller was sentenced to four years' imprisonment for drug smuggling in Dubai, after being discovered in possession of three poppy seeds, found on his clothing from a bread roll he had consumed in transit at Heathrow airport.[10]

A narcotic stimulant in Africa and the Middle East for thousands of years, khat is a legal stimulant and enjoyed on the streets of many European cities today. The leaves of the shrub are usually packed into the cheeks and masticated for their juice, like chewing tobacco. Khat is used by African Bushmen on arduous treks without sufficient food and water and modern-day Somalian insurgents, since its properties keep people alert and euphoric in testing circumstances. It grows wild in desert conditions, yet it is widely cultivated, as in Yemen, and can be harvested as many as four times per year.

Despite a recent reappraisal of the positive function of weeds in relation to our ecology in particular, they remain, for many, a *cache tout* notion of what is harmful. The farmer who refuses to see weeds as a potential threat puts his livelihood at risk. The idealistic and perhaps inexperienced gardener who wants to allow weeds equal rights and freedoms with other plants soon amends this lack of discrimination or faces a garden rampantly out of control and with a few dominant weeds taking over. Few experienced gardeners do not bear a grudge or two against some successful weed or other. What we think of as weeds, wilfully indifferent to our needs, have the potential to run riot like some apocalyptic triffid, spreading filth and disease, the fingers of their unseen subterranean roots causing subsidence under buildings and in some cases devastating existing habitats, just because they can.

This picture of greedy, malevolent – and therefore morally indifferent – vegetable growth stems from a sense of an idealized natural world that it is our duty to maintain, in relation to human needs and desires. Plants are seen as native or non-native, even though their origins are often impossible to trace with any degree of certainty – to

Marijuana, known as 'weed', became associated with political dissent in the 1960s.

before humans, to before the last ice age, challenging at what point a plant can be said to become native or alien. Many plant species have been introduced into Australia since European settlement. It has been argued that 40 per cent of the total flora in the British Isles was introduced from other parts of the world.[11] In the San Francisco Bay Area the rate of new plant invasion is said to have increased from a new species every 55 weeks from 1851 to 1960 to one every fourteen weeks from 1961 to 1995, with some suggestion that at this rate of change there is a risk that the 'earth's flora might eventually homogenize to only a few highly successful species'.[12] Perhaps a twenty-first-century Garden of Eden would include only well-behaved weeds, weeds that hardly merit the name, and the result would be nature as a managed theme-park garden, static and predictable and thus, arguably, reliant on human management.

Weed science is a relatively new discipline. Its birth can be traced to those who first began to observe the objective detail of plants that had been disregarded except in relation to the difficulties they caused our human objectives. John Stevens Henslow, mentor to both Leonard Jenyns and Charles Darwin, was one such naturalist:

Hypnos, the god of sleep, stands behind Endymion with a branch of poppies, pouring a sleeping potion over the beautiful young mortal, so that he can remain with Selene, the immortal goddess of the moon. Roman, marble sarcophagus panel, *c.* 210 CE.

> He would notice in his walks any little variation of character in the weeds by the road-side, that helped to confirm generally-received views respecting their organization and affinities, any monstrosities of leaves and flowers that served to establish the laws of vegetable morphology.[13]

In 1965, H. G. Baker drew up a list (updated in 1974) of weed characteristics that has formed the basis for much subsequent study, including both agrestals, weeds of agriculture, and ruderals, weeds of waste places and those that grow alongside roadways. He included plants that directly or indirectly injure or cause damage to crops; those that injure or cause damage to livestock, as in poisoning forage; plants which impede water flow or navigation; those that are dangerous to public health, causing pollen allergies or dermatitis, for example; and those that interfere with the environment, such as those leading to widespread change in the growth patterns of other species, causing major changes to ecosystems.[14]

Baker's 'ideal' weed has the following characteristics:

1. the ability to germinate in many environments;
2. discontinuous germination and seeds with a long life;
3. rapid growth from the vegetative to the flowering phase;
4. seed production that is often continuous, for as long as growing conditions permit;
5. they are frequently self-compatible, but not completely autogamous or apomictic. Plantain pussytoes, for instance, is a prolific weed in the east of North America, and reproduces either through seed or clonally through its stolons, which are stems that lie above the ground and root at the nodes, depending on its situation;
6. weeds are often available for cross-pollination, by visitors such as passing animals, insects or birds, or by the wind;
7. they produce high numbers of seed in favourable environmental circumstances;
8. they can adapt to produce seed in a wide range of environmental conditions, being tolerant and plastic;
9. they adapt to short- and long-range dispersal;
10. perennial weeds have vigorous vegetative reproduction or regenerate from fragments of root;
11. perennial weeds tend to be brittle and are thus difficult to pull from the ground;
12. weeds have the ability to compete inter-specifically by special means, such as rosettes, choking growth or allelo chemicals (the toxic chemicals produced by a plant to defend itself against her bivores or competing plants).[15]

However, while the list is a useful starting point, many highly successful weeds have few of these characteristics. Himalayan balsam, for example, possesses only two, despite its vigorous, invasive habit across the wasteland and gardens of northern Europe since the mid-nineteenth century. Some fairly harmless weeds, such as common speedwell and chickweed, score high. Many native and non-native plants have varying time lags before they come to 'exponential growth',[16] so that it is difficult to detect them before it is effectively too late for their easy removal. Crucially such a list of traits has not proved helpful in predicting which plants are set to become invasive.[17] The holy grail of the weed scientist is to make such predictions before a plant has become established – as hard to achieve as identifying a future mass murderer in their early childhood.

Weeds compete for light, water and nutrients with other plants, causing problems for both native plants and those grown as crops. For the farmer this can reduce land productivity. Crops can become infected with weeds and the cost of harvesting may be increased with binding weeds, for example, both impeding the use of hand scythe in less developed farms and damaging vast combine harvesters in more developed ones. Annual weeds dry out fast and can cause fire risk, and firefighting equipment can inadvertently pick up weed seeds and deposit them on other sites. Roadside visibility can be obstructed by overgrown weedy verges. On the other hand, weeds can host and provide pollen for numerous beneficial insects and provide food and shelter for wildlife.

Speedwell, coloured etching by M. Bouchard, 1774.

Little Weed's character lacks vigour in *The Flower Pot Men* (1952–).

I am drawn to the concept of a fickle nature, indifferent to our needs, which the continued success of weeds seems to epitomize. Weeds suggest what sudden floods and earthquakes demonstrate more dramatically: that despite our efforts the world remains beyond our control. Weeds might be said to remind us of the squalid as much as the beauty of the natural world, the rub between our artful organization of field and garden and the turbulent evolutions of the natural plant world.

The nature I grew up with was largely suburban. As a child I found it hard to distinguish between the character of Little Weed in *The Flower Pot Men* on BBC television and the weeds in our garden. And Little Weed then was not the articulate, sunflower earth mother of today's version of the programme, but a simple, modest weed, more like a dandelion, with the weakest of voices: all she could say was 'Weeeeeeed'. She needed my protection. If I picked a dandelion flower the white sap flowed and soon its head fell limp and the petals shrank together; I knew it should have been left alone to turn into a seed clock. My grandfather, on the other hand, would inject with poison any abhorrent dandelion or dock which dared to spoil his lawn. This left patches of crisp, dry leaves that seemed to reassure him, but by the following year they had mostly recovered and the tiresome process began all over again.

This same grandfather went by the nickname of Weed. Early in their marriage, my grandmother had him weed for her while she was heavily pregnant. He set up a series of mirrors so that she could observe and advise him on what was what from behind the French windows – thumbs up or down. He would fork a suspect out enthusiastically and lift it up for her to see, torn from its young and fragile roots. 'Weed?' he would mouth, and what could she do but smile wanly back, too late to save the life of some precious begonia or pansy seedling. Which brings to mind the old adage: if it comes out of the ground easily, it is a valuable plant – not a weed.

one

The Idea of Weeds

When we value a plant we do not call it a weed. A weed is not a wild flower or a herb. It should not be precious or sought after. It is said that what marks out a weed is our attitude to it. If a weed is called noxious or invasive or non-native, then it is being taken seriously. Weeds are often said to be 'common', a term bearing all the distaste of that epithet applied to our own kind, suggesting vulgar and unrefined breeding and behaviour. In her poem 'No Coward Soul is Mine' (1846), Emily Brontë described men's beliefs as 'worthless as withered weeds'. In a letter to Horace Walpole, the Countess of Ailesbury draws on an image of weeds to describe a lack of mental clarity in others: 'When people will not weed their own minds, they are apt to be overrun with nettles.'[1]

If you are a farmer, a highway verge manager or simply a domestic gardener, you must wage war on weeds. Weeds are the enemy and must be rooted out. Yet by their hardy nature they appear to fight back, for to be a weed is to have the ability to resist our attempts to destroy them. Plants vie with each other to survive, and perhaps because they seem less like us than members of the animal world, their battles slow, anchored to the ground, roots unseen and therefore potentially sinister, we ascribe to too-successful plants in the survival of the fittest the category of weed. Darwin wrote, 'It is difficult to believe in the dreadful but quiet war of organic beings going on in the peaceful woods and smiling fields.'[2]

Weeds of all sorts survive in apparently unfavourable conditions.

Yet this picture suggests that weeds return our animus, whereas all they do is succeed, quite oblivious of hopes and needs. We may inadvertently offer them shelter and new opportunities, and our weeding may even allow them to cover more ground more easily. We rip them out as seedlings, slice through and rotavate deeper root systems, deny them light or even drench their leaves with systemic chemicals. In response, despite our best efforts to remove all of an offending plant, tiny sections of root can easily become multiple new plants; seeds are carried to new sites where they can prosper anew; herbicides may not reach down deep enough to kill and may in fact strengthen a plant in the long term, toughening it up for future affray. Weeds, it might be said, are the success stories of the plant world, humans the success story of the animal kingdom, and we are set against each other. Who will be victorious?

Weeds, then, can be a concept for what we find threatening in the plant world, rather than a specific botanical or horticultural genus or species. The carrot with its taproot, for example, was once a wheatfield weed. The term 'weed' is used to describe our shifting attitudes. Kentucky bluegrass is welcome pasture fodder, but in Rocky Mountain National Park in Colorado it has become a threat

A weed hook and a crotch are being used to pick out weeds in this glass roundel, 'June', from a set depicting the labours of the months, *c.* 1450–75.

to native plant species, its rhizomes forming a thick mat of almost impenetrable growth. Elm trees, now highly valued given their rarity due to the ravages of Dutch elm disease across Europe and much of North America in the second half of the twentieth century, were once common enough in Britain to become known as the Weed of Worcester.

The idea of an ideal lawn depends upon its cultural context. A medieval lawn, before herbicides, would have been full of flowers and herbs: 'They were fragrant carpets to be walked, danced, sat and laid upon.'[3] More recently, moss in traditional, well-kept European bowling-green lawn grass had to be eradicated, whereas a Japanese verdant mossy sward must be painstakingly hand-weeded of all grass.

Burdock is a weed in the field but a prized plant in the border. Downy brome grass is considered a noxious weed on American prairie land, yet it can serve as a useful winter and early spring livestock feed.[4]

The weed is a category of plant that is traditionally associated with what is poorly designed. Aelius Nicon, the father of the physician Galen, studied seeds and came to believe that weed seed was merely a deviant form of well-bred seed. Fields were often sown with mixed crop seed, as *mesta*, and it was his theory that weeds such as darnel and *Aegilops* (*geniculata* or *ovata*, goatgrass, a noxious weed common in the Mediterranean) were mutations. When in the musical *My Fair Lady* Professor Higgins wants to get Eliza a gown suitable for Ascot, he insists, 'I despise those gowns with sort of weeds here and weeds there.' He may mean flowers in general, but the word he uses is weeds, to express his contempt for a design without any vestige of formal integrity.

Some weeds are considered to bear greater moral weight than others. They are dangerous, not only in their threat to selected, nurtured plants, but also in their very existence, which becomes a critical commentary on our human inability to remain in charge; aesthetically challenging in their so-called deformity and irregularity, they

Chinese women weeding the lawn of a wealthy foreigner in Shanghai, *c.* 1890–1923.

Burdock takes root in the Via Gellia, Matlock. Photograph by Francis Frith, *c.* 1850–70.

A study of burdock leaves by Carl Wilhelm Kolbe, 1826–8.

are also morally offensive, in what John Ruskin described as their stubbornness. He views weeds as a degenerate form of plant life and a metaphor for the human mind:

> drifted, helpless, entangled weed of castaway thought; nay, you will see that most men's minds are indeed little better than rough heath wilderness, neglected and stubborn, partly barren, partly overgrown with pestilent brakes, and venomous, wind-sown herbage of evil surmise . . . set fire to this; burn all the jungle . . . and then plough and sow.[5]

Though weeds are more often personified as tough thugs, in common parlance 'weedy' describes a person who is meek and lacking physical strength, like an etiolated plant, perhaps, lacking sufficient light and space to flourish. It is a term for an inferior horse, showing signs of a weak constitution. So a weed might be said to be a too-robust plant, like Audrey II, the man-eating pot plant in the rock musical *Little Shop of Horrors*, yet in other contexts pathetically ill-bred and unworthy of our attention. Weeds can host disease, strangle more refined growth and generally spread their unseen, deep-rooted contagion. Wild violets are too limp and their flowers are insipidly small, too prone to damp, dark corners, as if lacking upright *amour propre*; in contrast, dandelions are too lush and healthy, their vigorous, indestructible roots, gaudy flowers and too-plentiful seed heads all too easily spawned with their easygoing means of reproduction by parachute-like seeds, landing where they will, suggesting something of human sexual profligacy. In all the confidence of the nineteenth century J. C. Loudon decried weeds as 'mere savages, and botanical species as civilised beings'.[6] This suggests another aspect of these contradictions, for if the natural world should ideally be in our control, then a distinction between the natural wild and what is artificially selected and cultivated ceases to exist.

Some plants are characterized as weeds yet are held to be beneficial, increasing crop yield, for example, as with three weeds commonly distributed in dry areas of India to increase the growth of millet: Arabian primrose, buttonweed and cockscomb. In Mexico some weeds are deliberately left to grow, such as nasturtiums, because they have medicinal and ceremonial uses or again are thought to improve the quality of the soil.[7] In Australia *Echium plantagineum* is a noxious weed on grazing land, yet as dry forage it is valued as an emergency feed, so that it has earned the contradictory common names of both Paterson's Curse and salvation Jane. Despite the emphasis given to the dangers of introducing non-native species, in Argentina *Rosa rubiginosa* is deliberately used to resist overgrazing and to allow native species to regenerate.[8] Such instances are not exceptions but rather typical of the ways in which plants designated as weeds also can serve as useful non-weeds.

Despite recent attempts to embrace the wild and to avoid former prejudices in both agriculture and in municipal and domestic horticulture, the idea of a weed as an ill-bred miscreant still lingers. Plant life is indifferent to humanity, yet we ascribe intentions to plants when they seem to resist or even attack our interests. The conceit is so deeply entrenched that we can easily feel as if weeds wilfully obstruct us. The thistle patch that returns year after year despite our best

'Weeds Preferred to Wallflowers', postcard, *c.* 1910.

Violet from *De Historia Stirpium Commentarii Insignes* by Leonard Fuchs (1542).

efforts to remove it, or the ground ivy that seems to benefit from our attempts to pull it up, convince us that they mean business.

One might perhaps imagine plants having a will as more than a metaphor.[9] Weeds certainly display considerable ability to adapt and benefit from changing circumstance, which suggests intelligence and even cunning. The American journalist Michael Pollan suggests that consciousness might not be all that important in the survival of the fittest, asking us to consider whether many plants are not successfully outplaying us in that respect. Agriculture legislates against unwanted weeds, but in fact, in clearing forests for land, we allow weeds further opportunities to grow and prosper. One might even imagine that grass, for example, got us to clear forest so that it had better opportunities to grow. When we mow grass we strengthen its roots, allowing it to out-compete trees. We talk of invasive weed species destroying a natural habitat, for instance, yet how did that habitat come about? Plants thrive and fail dependent upon fluctuations in the weather, on the consequences of great ice age shifts, on overwhelming deforestations brought about by mankind and even on the minutiae of small, local

A nasturtium in pencil and watercolour, by Toni Hayden.

changes in disease and animal husbandry. The consequences of Darwin's theory of natural selection tend to be ignored in any conventional view of weeds.

Unlike much managed planting and agriculture, weeds often grow in mixed communities, which have given them the resistance to disease that monocultures can lack. When necessary to their survival they may be capable of developing asexual reproduction, and their methods of reproduction may adapt to different environments. In some instances they are able to repeat several life cycles in a season when there is sufficient rainfall. Their seed may be buried for many years until conditions improve. The oldest carbon-dated seed that has grown into a viable plant was a Judean date palm seed, about 2,000 years old, taken from an excavation at Masada in 1973.[10] Pollan describes the 'thousands of weed seeds [which] lie dormant in every cubic foot of soil, patiently waiting for just the right combination of light and moisture'.[11] Instead of relying on human intervention, seeds may spread via bird and animal dung, be carried by insects, rain and wind, be borne across the oceans on the tide, or travel in packing material for porcelain imports, as Japanese stiltgrass does, or even cadge a lift in the trouser turn-ups of a learned botanist. And as Charles Darwin's study of his weed patch at Down House in Kent demonstrated, as we shall see, the hardiest plants will gradually take over a cleared piece of ground, supporting his theory of the survival of the fittest.

Weeds and other plants are often interdependent in ways that are only discovered when a weeding programme is too successful. A parallel might be drawn with Mao Zedong's attempts to rid China of sparrows in order to protect grain crops. People were ordered to bang on woks and pans to keep the birds from resting until they dropped exhausted from the sky. However, in consequence caterpillars, free from predators, infested their crops, which led to widespread famine – until gradually the sparrows returned. The relationship between plants we desire and those we do not at any one time is complex. Since weeds thrive in warmer, damper climates more easily than in colder, drier ones, it might follow that tropical regions are at greater risk and that global warming means the threat is increasing. A plant considered a weed in a tropical clime may come to grow in more temperate regions because of global warming, but initially, its rarity may mean that its potential to be a weed is not recognized.

Some weeds are dependent on agriculture, agrestals requiring 'human management to maintain their populations'.[12] Moreover,

Harvest time, with tares being burned in the background and weeds pulled in the foreground. Print by Pieter Jalhea Furnius after Gerard Groenning, 1585.

when we disturb the soil, and the environment more generally, for any purpose, some weeds take the opportunity to take root.

A weed solution can lead to unforeseen new problems. A plant we may find unsightly might be the very one on which other plants rely in order to thrive. Its removal can create opportunities for other more noxious plants to prosper.

two

The Background

In happiness
or sadness
weeds grow and grow
SANTOKA TANEDA, 1882–1940

Imagine for a moment a world without weeds. It follows that this would have to be a time before people, for without language such abstract ideas have no means of existing. Birds, grazing animals and insects came about and lived among vegetation in prehistoric times. The group of plants we now think of as weeds must then have impeded the growth of other less robust plant life. Some would have been abundant, welcome feed and others would have poisoned animals. They may have had no generic names, no names at all in fact, but it makes little sense to imagine that plants with the potential to be called weeds did not exist.

The early forerunners of the plants that are now deemed weeds grew and prospered. Yarrow, for example, grew in what is now Iraq over 60,000 years ago, its pollen discovered in Neanderthal burial caves. Wild emmer seeds were found in Neolithic settlements in Southeast Asia. Familiar common or garden weeds played their part in the long history of magical beliefs and herbal medicine. The ancestors of the dandelion, nettle, chickweed and poison ivy provided poultices for wounds, cures and palliatives that were believed to alleviate desire, purify the blood and improve the complexion,

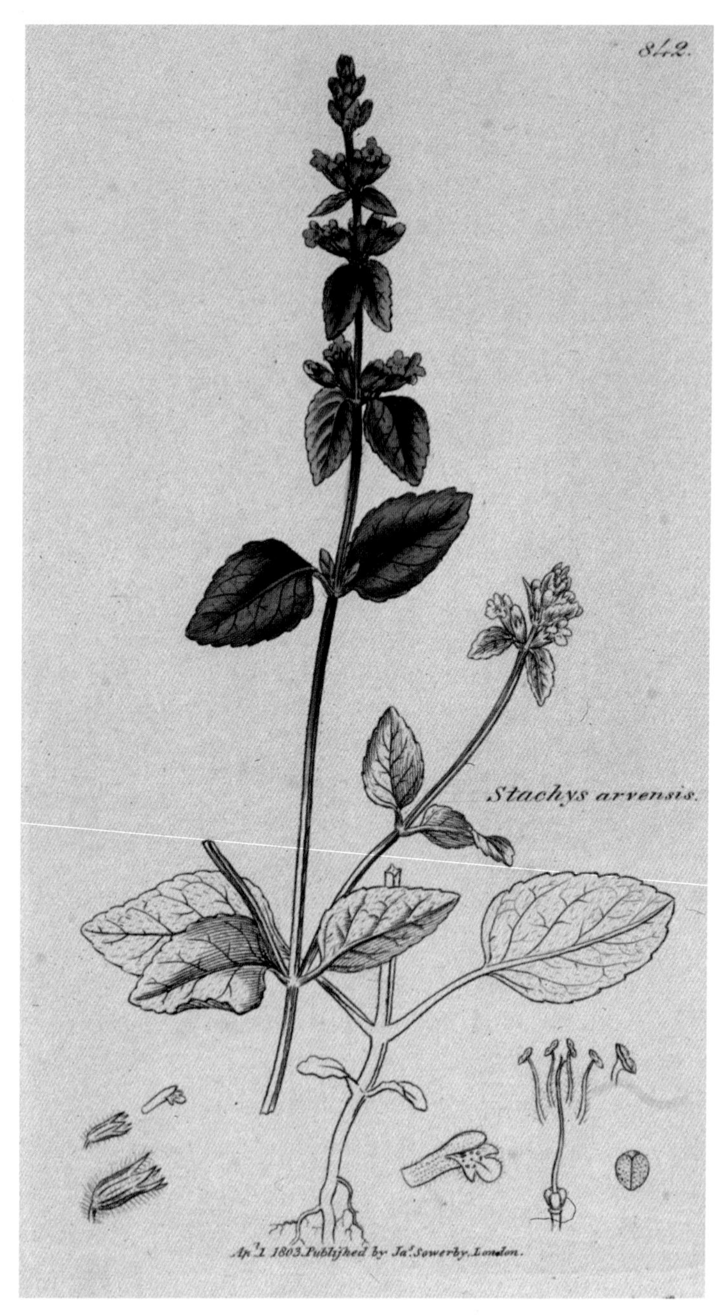

A hedge nettle or woundwort, showing the flowering stem, roots and floral segments. Coloured engraving by James Sowerby, 1803.

and seemed to some to exist solely to sort out these and many other human difficulties.

The first time early man stigmatized one plant as opposed to another, when it was found troublesome or unlovely, weeds as a concept came into being. There is evidence in cave paintings and in archaeological finds of weeding and weeding tools. Neanderthal man has been shown not to have kept to a narrow, meat-only diet, as was once believed, but to have eaten wild plants, including grass species, the remains of seeds and vegetable residue having been found on tools and calcified on teeth. These plants were sometimes cooked and sometimes processed and ground to make them more palatable. Hunter-gatherers selected what was edible or otherwise useful to them.

The first farmers quickly learned the need to protect their crops by weeding out unwanted, more vigorous plants. Even today the image of the farmer or gardener usually suggests someone working to keep weeds at bay, with primitive hand tools such as the hoe or grubber. There are many engravings and wall hangings of Chinese farm workers at the never-ending business of weeding rice paddies. Similarly, there are many ancient Japanese scrolls depicting such weeding, often a solitary female, back bent over her task. From as early as the fourteenth century the English illuminated manuscript called the Luttrell Psalter shows weed hooks and crotches being used to pick out weeds from a field.

Some plants now considered noxious weeds were once valued and even domesticated for their culinary and medicinal properties. Plantains, teasels and wild carrot, for example, were listed as useful medicines in Dioscorides' *Materia Medica* of 50–70 CE. Nutsedge is now widely considered one of the most troublesome weeds in temperate agriculture, but it was once widely valued for its tubers, or tiger nuts. Theophrastus described tiger nuts, or *mansion*, boiled in a barley beer until sweet in his *Historia Plantarus* in around 300 BCE, and these sweetmeats were also referred to by Pliny the Elder and enjoyed by both Mycenaeans and Assyrians. It is one of the most ancient Egyptian foodstuffs we know of, with tomb paintings from the fifteenth century BCE depicting recipes for tiger nuts sweetened with honey, and tubers have been found in tombs from pre-dynastic times to the Roman period. Red poppy flowers and blue cornflowers were inlaid on a box in Tutankhamun's tomb and his sandals were decorated with what appear to be mayweed flowers. The berries of *Withania* nightshade, which grows in waste areas in northern and tropical Africa and across Asia, were also discovered in Egyptian tombs, threaded onto thin strips of date palm leaf, a practice that was still current 1,000 years later in the Graeco-Roman period.[1] Pliny referred to nightshade being used in Egyptian garlands; its fruits were known to have powerful medicinal properties. The poet Theocritus of Syracuse mentioned more than 87 different types of plant, and his interest in wild plants was said to be 'without precedent in the whole range of early Greek literature'.[2]

In the Old Testament thorns and thistles are a punishment for humans' failure in the Garden of Eden, a curse from God. Man's expulsion from paradise meant that plants, instead of simply answering our needs, could become our enemies. We would be forced to live by the sweat of our brow, forever trying to weed out, to undo what can never be undone, for

> cursed is the ground for thy sake; in sorrow shalt thou eat of it all the days of thy life;
>
> Thorns also and thistles shall it bring forth to thee; and thou shalt eat the herb of the field;
>
> In the sweat of thy face shalt thou eat bread, till thou return unto the ground; for out of it wast thou taken: for dust thou art, and unto dust shalt thou return. (Genesis 3:17–19)

We are dry soil, infertile soil, incapable of nurturing even a wild weed. One may wish to imagine the perfect garden before Adam and Eve sinned, and it seems strangely at odds with many of our notions

The beauty of bramble, borage and goosegrass in a midsummer hedgerow.

of what is valuable about the natural world. Eden cannot be man in control of plants, so that they behave themselves in relation to him, for it is God who must have been cast in the role of the all-powerful gardener. If it is a perfect consciousness that creates each and every plant, the notion of weeds – of unwanted weeds – makes little sense. Thus, as with the more general problem about an ideal nature, field or garden without weeds, nature loses its wayward, invigorating delight, which is often what we most like about it. At any rate, such a garden had surely to be a form of *hortus conclusus*, an enclosed space, presumably with wilderness, wild and weedy, beyond, or at least its possibility. For there to be a paradise, there must have been some contrasting state, even before the Fall.

The many versions of a perfect and yet natural garden, actual and mystical, continued to hold force right until the Enlightenment, when the authority of the Bible might be said to have begun to falter. Yet as an ideal it continues to find its modern-day expression in ideas such as the eco-garden, as discussed in chapter Seven. Weeds also appear as a metaphor for man's failure, his cities destroyed and overrun by wilderness, where thorns and stinging plants contrast

Madonna and Child with Saints in the Enclosed Garden, oil on panel, *c.* 1440–60.

with notions of the paradise garden: 'And thorns shall come up in her palaces, nettles and brambles in the fortresses thereof' (Isaiah 34:13).

Medieval Arabic philosophy offers images of the weed – in Arabic *al-nawābit* – as a parable of the politically active. Al-Farabi (*c.* 872–950) drew a parallel between their growth, their steadfast well-groundedness and the role of rigorous truth-seekers.[3] Ibn Bajjah, or Arempace (*c.* 1085–1136), on the other hand, claimed that 'one of the characteristics of the virtuous city is that it is free of weeds.'[4] While Al-Farabi described weeds as pushing up into the light from fine lawns, Ibn Bajjah, in contrast, had them growing in coarse crabgrass, suggesting that someone would need to be similarly tough to enter the jungle of political life. Even with Al-Farabi's use of the image, weeds do not always strive for truth. As with more general attitudes to weeds, they are a slippery concept here, and the wise ruler has to keep a careful eye on them in case, through error or perfidy, they begin to cause trouble, just as we need to keep an eye on weeds, even the seemingly well-behaved ones, lest they become too rampant. They remain pragmatic rather than righteous.

Mary, Queen of Scots, whiling away her long period of imprisonment, worked on embroideries with Elizabeth (Bess) Talbot, Countess of Shrewsbury. One of the most beautiful surviving pieces, part of the Oxburgh Hangings, shows what appears to be a simple dandelion, with silk and silver-gilt thread worked on linen canvas mounted on silk velvet. One might imagine that such a weed, a symbol of irrepressible natural force, might have seemed to be appropriate for that beleaguered woman.

The Arts and Crafts Movement was wedded to the idea of elevating what was dismissed as too humble for notice, often incorporating weeds into their designs, such as in Charles Voysey's 'The Furrow', 1902–3, with a motif of common daisies and dandelions, and in Morris & Co.'s 'Woodland Weeds' of about 1905, both produced as wallpapers. William Morris's view that 'the Middle Ages was a period of greatness in the art of the common people' supported the idea that all unsophisticated peasant labour was valuable of itself, so that even weeding was to be valued aesthetically. The influence of the movement spread across Europe and North America and also

A dandelion embroidered in silk and silver-gilt thread on silk velvet by Mary, Queen of Scots, 1570–85.

Japanese ceramic sake bottle with delicate water weed motif.

further afield as, for example, in the Japanese Mingei movement of the 1920s and '30s, which celebrated artisanal Japanese, Korean and Chinese pottery, often decorated with simple weed designs, said to be 'beyond beauty and ugliness'.

The appreciation of humbler plants might be traced to the history of flower arranging. Consider Japanese ikebana flower arranging, its interest in simple line and form, with often the most modest wild weed plants used, linked to a Buddhist idea of returning to a nature without the intrusion of human organization. The Japanese tea ceremony should include a flower arrangement that suggests the participants are out in nature. In contrast there is the luxuriance of summer roses, spring hybrid tulips and exotic fruit in deliberate displays of affluence in Northern European Renaissance arrangements, allied to exotic bulbs coming into the Low Countries from Turkey as early as the sixteenth century. Victorian bunches crammed together all the abundance of the nineteenth-century bourgeois garden.

The Englishwoman Constance Spry was a modern florist, one of the first women to dominate the profession, who used unusual materials including weeds and grasses. Working as a head teacher in a poor part of east London, she discovered that her pupils seemed to enjoy learning to arrange the foraged plants she brought in for them. Flower arranging had long been a pursuit of the rich alone, who could afford cut flowers. Spry showed her students that they could gather plants for free. In time she was to become a fashionable society florist, and her display of hedgerow plants, which filled the shop window of Atkinsons perfumery in London's Old Bond Street in 1927, made her reputation. She was employed by royalty and at the high-society wedding in 1938 of Jo Grimond and Laura Bonham-Carter she set up huge urns not of lilies and orchids but of cow parsley alone, all along the aisle of St Margaret's Church, Westminster. This new fashion for 'country flowers' was seen as revolutionary after the constraints of Edwardian 'carnations and asparagus fern' arrangements, but there remains a connection to the restraint and modesty and also the careful artfulness of ikebana. Spry claimed that her tastes had merely

'The Furrow' wallpaper with flowering daisies and dandelions. Colour machine print designed by Charles Voysey, 1902–3.

been lucky enough to connect with the zeitgeist, when 'people were getting tired of the conventional set-pieces made by professional florists and were not entirely satisfied with purely amateur arrangements.' When the Design Museum in London held a retrospective exhibition in 2004, it met with considerable critical uncertainty. A journalist attempting to defend Spry's work suggested that 'The starting point of her philosophy was that wild flowers and weeds could be pressed into service, as much as tuberoses. One could indeed spend a fortune. One could also spend next to nothing.'[5]

Spry's close friend the artist Hannah Gluckstein, known as Gluck, was influenced to include wild flowers and weeds in her flower paintings, such as *Dead Flowers*, which includes cow parsley and poppy heads. Today there is again a trend for using weeds in flower arranging, perhaps a reaction to the use of exotic blooms like tropical

'Woodland Weeds', a wallpaper design by John Henry Dearle for Morris & Co., *c.* 1905.

Cow parsley and thistle textile design by G. P. and J. Baker, 1900.

parrot flowers and orchids. The country bunch idea has been seen in many fashionable weddings. Kate Middleton's bouquet, unlike Diana Spencer's more elaborate one, was small and included common flowers and ivy, and many other bridal bouquets use field flowers such as cornflowers and poppies, cow parsley, purple deadnettle, dried grasses and even sea cabbage, as if there were a connection between their plant ordinariness and a bride's modesty.

In fashion, a liking for weeds tends to mark a turning away from more conventional choices. The end of the nineteenth century brought with the Pre-Raphaelites a new way of dressing in the bohemian style, engaged with the idea of the natural wilderness: drooping robes belying the corsetry beneath, men in open-toe sandals, open-necked shirts and unkempt beards. A weed-like fashion can be expressed through textiles, such as in a preference for green and generally natural colour schemes, and for floral print on bucolic dress, alternative hippy and ethnic garments. Such clothes are often loose and diaphanous, with gendered distinctions often deliberately confused, suggesting sexual licence; hair was worn long by both sexes or women were shorn, with bare feet and a lack of apparent careful grooming.

Physical languor and relaxed 'chilled out' deportment are important. These features might be compared to the idea we have of untidy, sprawling, misbehaving weeds. Laura Ashley's designs of the late 1960s–70s commercialized this look, from flower child to milkmaid, appearing to eschew a desire for status. I recall a shaggy, layered haircut in the early 1980s, known as the *coupe sauvage*, intended to make one feel both fashionable and somehow beyond fashion. It means to look untended and wild, as if you had been dragged through a bush. One might say that flowers are a little like aristocrats or the bourgeois, whereas weeds are somehow more working-class.[6] Today there remains

The Rosebud Garden of Girls, photograph by Julia Margaret Cameron, 1868. The girls' long, loose hair and flowing peasant dress are in tune with wild flowers and tangled weeds.

Cow parsley with bladderseed and chervil, after William Dickes and Anne Pratt.

an association between weeds and political dissent: a weedy garden may suggest non-conformity, so that one Internet blog following street fashion is called Urban Weeds. It is meant to recall the historic term for clothing, no doubt, but there also lingers a more edgy reference to urban cool, disrespectful, alternative, relaxed I-don't-give-a-damn *élan*.

In Flanders Fields in the First World War the wild corn poppy became a symbol of survival, of beauty and resilience. The poppy grows in disturbed soil, and so thrives in arable land and on roadsides and in waste places, both agrestal and ruderal. Its seeds can travel up to 3 m (10 ft), though may go further with the help of windy weather and birds carrying the seeds in their droppings. Like those of many successful weeds, these seeds are capable of lying dormant for many years before circumstances change so that they can germinate. These attributes made them peculiarly successful in the muddy desolation of the trenches, where often they were the only plants still managing

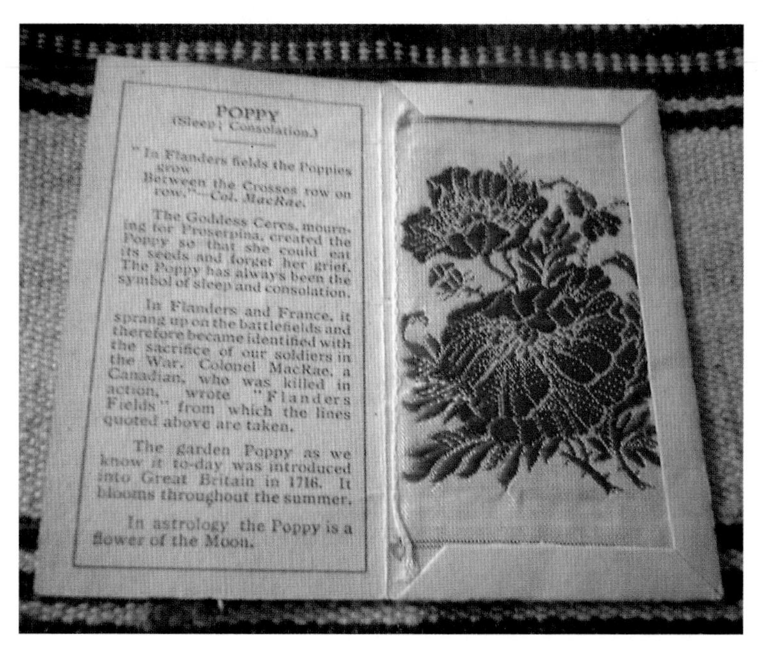

Embroidered silk poppies from inside a cigarette packet sent out to Flanders in 1914.

to grow and flower. On the one hand they brought the relief of beauty in terrible circumstances, and on the other they seemed to suggest the numbers of young men killed, their apparently cheerful colour aping the blood that was shed. John McCrae's poem has them 'between the crosses row on row', like the dead perhaps, rising from their graves. The young flower bud droops on its stem as if reluctant, then the crumpled scarlet petals unfurl. The flowers are short-lived, lasting only a day – and it is this habit and fragility perhaps that made the wild corn poppy so lasting an image of the entrenched soldier.

Archaeologists have discovered wild emmer seed in Mesopotamian and Roman sites. Ground elder, for example, was introduced into Britain by the Romans to relieve the symptoms of gout. In ancient societies, and in foraging aboriginal cultures that survive today, wild plants or weeds are sought after both as medicine and for the pot. Native Americans called plantains, brought over to the New World, 'white man's footsteps', so closely did they follow in the path of the settlers. The term was also used for nettles. In Crete wild chicory, *stamnagathi*, is gathered every Easter, and similarly in France there is the annual foraging tradition of *la cueillette*.

In nineteenth-century America, Henry Thoreau, influenced by Ralph Waldo Emerson and his notion that weeds were merely plants in the wrong place, advocated gathering wild food. Indigenous plants seemed to him to be more valid than imported varieties. He took great pleasure in the vivacity of willow staves, cut down to support a neighbour's bean crop, setting root and growing more strongly than the crop itself. This ability to survive impressed him, like the new settlers managing to survive the privations of a new continent. He reckoned that a Latin name granted them more respect:

> See these weeds, which have been hoed at by a million farmers all spring and summer, and yet have prevailed, and just now come out triumphant over all the lanes, pastures, fields, and gardens, such is their vigor. We have insulted them with low names, too, – as Pigweed, Wormwood, Chickweed,

> Shad-Blossom. They have brave names, too, – Ambrosia, Stellaria, Amelanchia, Amaranth, etc.[7]

Nonetheless, Thoreau's romantic view of nature was challenged by his attempts at growing beans, for to grow a crop is to impose a desired order on nature. He had to 'wage a long and decidedly uncharacteristic war, "filling up trenches with the weedy dead"'.[8] Michael Pollan describes this less romantic viewpoint as being short-lived. Thoreau is only playing at farming, after all, and soon he reverted to Emerson's metaphysical position:

> The sun looks on our cultivated fields and on the prairies and forests without distinction . . . do [these beans] not grow for woodchucks partly? . . . How can our harvest fail? Shall I not rejoice also at the abundance of weeds whose seeds are the granary of the birds?[9]

One can only guess that had he been dependent upon his crop for his living he might not have returned to such a benign view of the natural order.

Across the globe, foraging for freely available wild crops, on the periphery of cultivated fields or on any available common land, still survives – sometimes out of need but often, in the West, as a way of reconnecting, like Thoreau, with some imagined romantic past.

Cultivated land tends to involve the leaving of bare soil between the planting. Out of season, when the crops have been removed, the land is thus left open to the elements, and can easily become eroded, necessitating mulching and the use of fertilizers. Disturbed soil invites the wayward wind-blown weed seed that would never have found purchase in land left to its natural devices. As ploughs became more efficient, the damaging effects on the earth become greater. While gardeners like Gertrude Jekyll and William Robinson might praise in-filled planting schemes, with annuals and biennials together in mixed borders, farm crops are weeded and sprayed to create a

A vase of cultivated flowers, which bloom at different times of the year; still-life by Gustave Courbet, 1862.

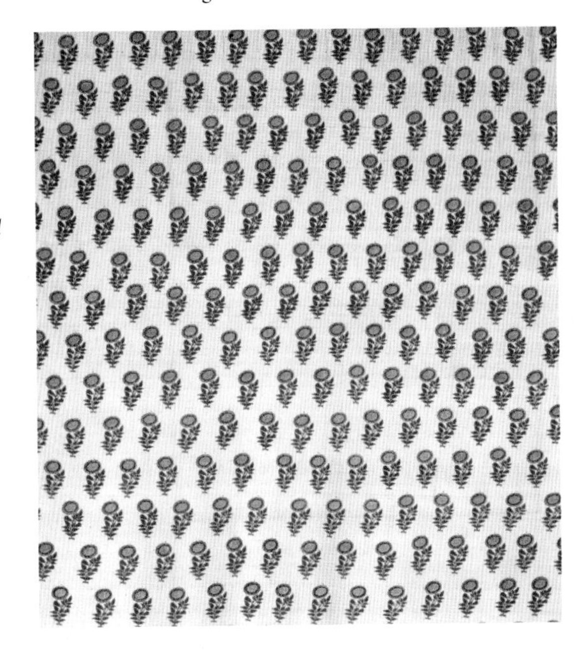

Block-printed silk fabric with a motif of sunflowers or marigolds, by Thomas Wardle, 1878. Taken from J. C. Robinson's *Treasury of Ornamental Art* (1857), in turn derived from a Persian design.

monoculture, which is traditionally thought to avoid disease and encourage uniformity in the yield. The history of the lawn, from naturally occurring meadowland to green sward packed with a multitude of species, like Giovanni Boccaccio's 'green grass powdered with a thousand flowers', from Roman camomile and medieval cloisters to Victorian and Edwardian close-cropped grass-only perfection, and contemporary American suburban lawns, all about keeping your front yard in order – all might be said to reflect this development from multiplicity to uniformity. The plant-naming system applied by the Swedish naturalist Carl Linnaeus in the eighteenth century, on which we still rely, demonstrates the elastic dichotomies between welcome plants and what are often staunchly considered to be weeds.

The world was gradually evolving with inherited qualities passed down from parent plants, according to Darwin's theory of natural selection, which had itself depended on close observation of weed species. His ideas seemed to entail the 'destruction of all the less hardy ones and the preservation of accidental hardy seedlings'.[10] When Darwin moved to the country, to Down House in Kent, he

Bare soil between plants is considered a sign of good husbandry.

set up an experiment in the grounds to test his theory. Digging up an area of his orchard, he left it bare and surrounded it by a fence, and then watched and waited to see what might happen to grow there. It was weeds that grew, of course, and to each new seedling he attached a wire, in order to plot each individual's progress. He marked the ruthlessness of the selection progress as coarser weeds drove out the more delicate ones:

> I have been interested in my 'weed garden' of 32 square feet [3 sq. m]: I mark each seedling as it appears, & I am astonished at the number that come up & still more at number killed by slugs etc. – Already 59 have been so killed; I expected a good many, but I had fancied that this was a less potent check than it seems to be; and I attributed almost exclusively to mere choking the destruction of seedlings.[11]

The majority of the weed seedlings failed to survive:

> My observations, though on so infinitely a small scale, on the struggle for existence, begin to make me see a little clearer how the fight goes on: out of 16 kinds of seed sown on my meadow, 15 have germinated, but now they are perishing at such a rate that I doubt whether more than one will flower. Here we have choking, which has taken place likewise on great scale with plant not seedlings in a bit of my lawn allowed to grow up . . . I have daily marked each seedling weed as it has appeared during March, April & May, and 357 have come up, & of these 277 have *already* been killed chiefly by slugs.[12]

Darwin's theory failed to explain the way in which some weeds can quickly adapt to changed circumstances, this ability being in part what causes them to be considered weeds in the first place.[13] It took Gregor Mendel's theory of inheritance, which allowed for sexual reproduction to introduce random characteristics to plants, to account for the way in which some are carried in discrete units, or genes, from generation to generation.[14]

It is the enclosure of land for agriculture in the West, at the end of the eighteenth century and into the nineteenth, that might be said to have established the notion of the weed. Neolithic farmers began the process of selecting plant species, and the agricultural revolution made it possible to restrict land use, segregating desired crops from all other native species, which thus became weeds.

The poet John Clare (1793–1864) railed against the changes brought about by agrarian capitalism and enclosure to his own childhood home in the Northamptonshire village of Helpston. He considered modern farming a personal tragedy, like the expulsion from Eden itself. In his poem 'Remembrances' he described how his 'boyhood's pleasing haunts' were 'shrivelled to a withered weed and trampled down and done'. Once the soil was tilled and plants were sown, competing plants became a threat and had to be eradicated or at least subdued. Employed as an agricultural weeder just as the agricultural revolution in England was getting under way, Clare's observations appreciate both the beauty of his surroundings but also the miseries of such manual labour. Rudyard Kipling honoured the manual labourer in his 'The Glory of the Garden' (1911): '. . . better men than we go out and start their working lives / At grubbing weeds from gravel-paths with broken dinner-knives.'

❧

From the ordinances of Henry II in the twelfth century to the Weeds Act of 1959 in Britain, from Noxious Weeds Acts in Canada and many American states, in New Zealand from 1908, 1950 and 1978, and the immensely detailed legislation of the South African law of 1983, to the many attempts to stop the constant invasion of the Australian ecology – in societies where land has been cultivated, those in power have attempted to legislate against weeds, mimicking perhaps the efforts of a King Cnut. In the nineteenth century, rapid advances in technology, attempts to manage the natural world and in particular

A forest in Uganda snowed under by the white blossom of knotweed, 1936.

food production methods led to a counter-reaction of nostalgic ideas about the sanctity and romance of wild nature, which in turn brought about a reappraisal of how weeds were seen.

The upheaval of the agricultural revolution was followed by the widespread deforestations and ecological upheavals of the Industrial Revolution, with land enclosed for increased food production and many moving to the cities in search of new factory jobs. Before the 1790s, for example, over 90 per cent of Americans worked on farms. In the 1830s the steel plough, invented by John Deere, and Cyrus McCormick's steel reaper, meant that increasing expanses of grassland were being turned into arable land, meaning more food crops but fewer workers required to farm it. Ploughing was in one respect the best and easiest means of weeding land, but it also meant that turned, plant-free soil could more easily become infested with quick-growing weeds and small sections of roots of tougher weeds left behind to grow and multiply. Many plants were driven to extinction, but the stronger, invasive weeds benefited.

Japanese knotweed taking over a west London garden in 2014.

For some this rapid change in our ecology is seen as evidence of hubris in relation to the true natural order. This response might be considered as a sense of nostalgia for the past. Just as Clare had longed for the country of his youth, William Cobbett also regretted the changes to the countryside of *his* youth of 30-odd years earlier, the very England that Clare glorified.

The rhododendron that has become an invasive plant, *Rhododendron ponticum*, was introduced by Conrad Loddiges into Britain from the Mediterranean in the late eighteenth century, and became a fashionable ornamental plant. After several harsh winters in the late 1800s, which knocked back native British species, without competition it gained ground. The glossy-leaved shrub with its brilliant cerise flowers began to seem not so desirable after all. It became brazen, a foreign weed.

Similarly, Japanese knotweed is an attractive ornamental plant in the thin, volcanic soil of Japan but has galloped across European loam. Vast sums of money were needed to clear the London 2012 Olympic site, for example, from its encroaching, foundation-wrenching root system, which can reach to a depth of 7 m (23 ft) and can regenerate from as little as a 1-inch length from any part of the plant. A company set up to eradicate the plant describes its technique as involving 'excavation, dig and dump, on-site burial, soil-sifting and our own innovative stem injection system'.[15] The plant was introduced into Europe in the mid-nineteenth century from Japan and eastern China. Giant hogweed came to western Europe from the Russian steppes.

Weeds become a serious threat in new conditions. Australia has suffered from the proliferation of many imported plants that caused little problem in Europe, such as privet and laurel, but which have become environmental disasters there. The Scottish thistle was brought over by early settlers and now is grown out of control. The parasitic wild snapdragon, a plant strewn in the path of honoured visitors in Kenya, has become a severe threat to crops there, due to modern spraying destroying their indigenous competition. In the Vietnam War, Agent Orange not only defoliated the jungle but also allowed cogon grass, a native plant that was previously inhibited by the shade of the jungle canopy, to stage a takeover.

When crop plants mutate, their feral offspring may be considered weeds. Their continued close relation to crop plants means that they may be hard to control, with a built-in resistance to herbicides designed to leave crops unaffected. Such feral plants may be the outcome of 'crop seed that shattered prior to harvest in previous seasons'.[16] Yet what might be seen as a Darwinian process of plants transmuting and vying with each other to survive is something experts in the field have to find a way of controlling. These wayward offspring are not only capable of infiltrating agriculture but 'may enter the ecosystem of wild species', where with further cross-breeding they may return again to threaten crop yield. One might say that the tearaway goes native, lies low and then returns to wreak havoc all over again.

Ragwort is widespread: both pernicious weed and essential insect food. It thrives on wasteland and the margins of roads.

An example of plant proliferation that saw weeds in a largely more positive role is that of the weeds that grew in the London Blitz during the Second World War. In the Great Fire of London in 1666, a similar abundance of weeds took over the burned and cleared ground of the old city. The extraordinary flowering of wild plants on London's excavated bombsites created a new ecosystem, which included Oxford ragwort from Sicily, Peruvian gallant-soldier and rosebay willowherb – christened bombweed or fireweed – along with buttercups, chickweed, nettles, dock, groundsel and plantain.

One hundred and twenty-six species in all were logged by Edward Salisbury, director of Kew Gardens towards the end of the Second World War, who described them as having taken root in the London's wounds.[17] Suddenly light flooded into what had been dark, dank cellars or narrow alleyways. Wasteland flourished with seeds blown in, or buried deep in the earth many years before and now exhumed.

Rosebay willowherb is an early colonizer of disturbed ground, sometimes termed a pioneer species, and tends to lose advantage as new buildings, forest canopies or vegetation grow up nearby. In the eighteenth century rosebay was considered a rare woodland plant in Britain. On the Isle of Wight rosebay was recorded in 1909 as having flourished in woodland that had been destroyed by five. Today, while it is successful on set-aside land, as soon as annual crops are planted it is less of a problem. Once established among perennial planting it can be much more difficult to eradicate. Where there is little or no competition, colonies can survive for many years, with stands of rosebay surviving for 35 years on Dutch sand dunes, for example. It is unusual in that its leaves and flowers smell quite different from each other; the almond scent of the latter is said to have been a favourite of Elizabeth I of England, and was used to perfume her bedchamber. In his *Herbal* (1597) the botanist John Gerard (*c.* 1545–1613) wrote:

> The leaves and floures of Meadowsweet farre excelle all other strowing herbs for to decke up houses, to strawe in chambers, halls and banqueting-houses in the summer-time, for the smell thereof makes the heart merrie and joyful and delighteth the senses.

Pliny also praised rosebay willowherb for its beauty. Gerard described it as:

> growing to a height of six foot, garnished with brave floures of great beauty, consisting of foure leaves a piece, of an orient

> purple colour. The cod is long . . . and full of downy matter which flieth away with the winde when the cod is opened.[18]

A strain brought over from Europe to America in the early 1800s grew in such profusion that it threatened to clog waterways from the Hudson through to Alaska. The indigenous rosebay was a food source for Native Americans, who ate the younger shoots raw, the older stems peeled and roasted. The Eskimo word *pahmeyuktuk* is a term for the edible shoots soaked in seal oil, a useful food source of vitamins A and C. Its sap is used in sweets and syrups in the northwest of America; it is also an important source of nectar for honeybees. Along with goosegrass or bedstraw, rosebay is the preferred food of elephant hawk moths. In Russia, rosebay is a tea substitute and in Austria the tea is taken to ease urinary complaints.

Weeds in their infinite variety can help and hinder in this way. Rosebay's apparently shy nature, given its habit of making way for other plants, is set against its more brutish ability to spread and grow fast up to 2 m (6 ft 6 in.) and sometimes even more. It successfully followed the path of many northern hemisphere railways as they were built during the nineteenth century. Its fluffy seeds are easily driven by the wind to find safe purchase, and if they do not, they have the ability to remain viable deep down in the seed bank until many years later, when fires, ploughs or other disruption might allow them to germinate. As a garden weed its roots are easily snapped if uprooted, small segments are able to sprout and form new plants, and they can quickly send shoots under paving slabs or grow out of cracks in walls. Yet with a little work it is easy to control, as long as the seedlings are rooted up every year. A beautiful name and plant, but like every weed, like every plant, its aim is to survive above all.

It is an irony of nature that out of deliberate destruction comes, as an unforeseen consequence, a healthy growth of weeds. One instance was the fast weed colonization of Hiroshima. In John Hersey's *Hiroshima* (1946), Miss Sasaki is on her way to hospital and sees the city for the first time since the atom bomb had been dropped. Over 70,000

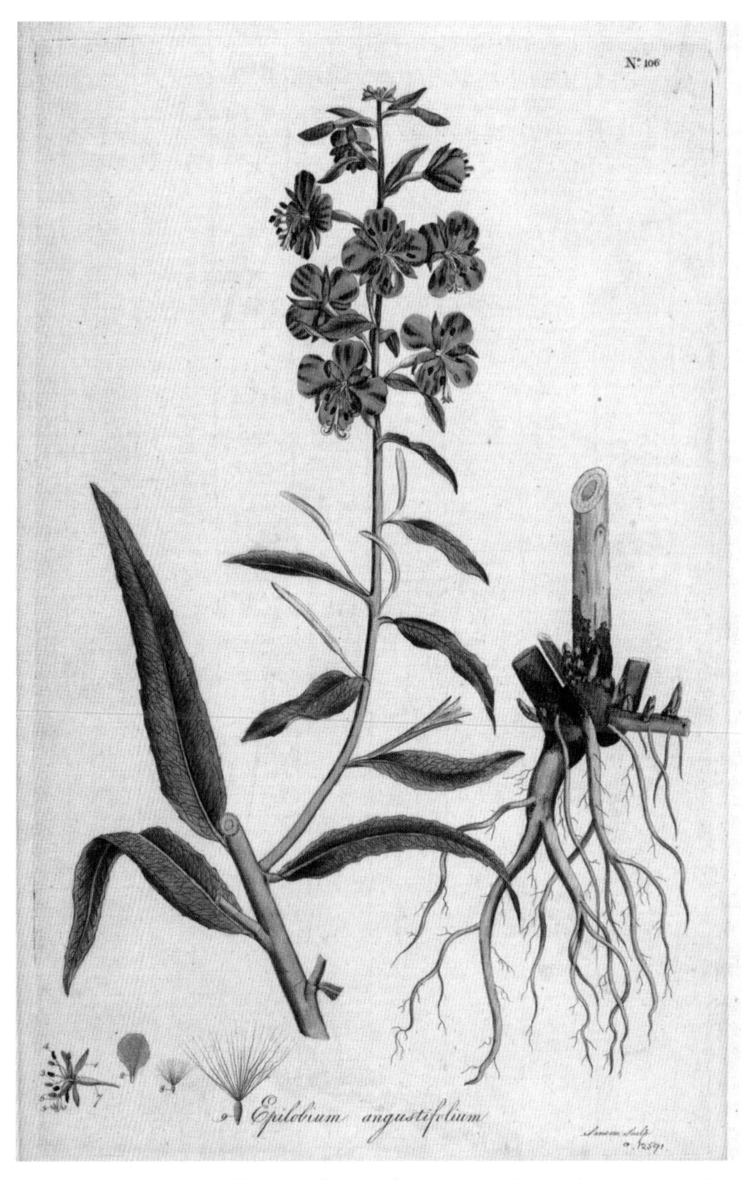

Engraving of rosebay willowherb, from William Curtis's *Flora Londinensis* (1777–98), a field guide to the wild flowers of London.

people had been killed and as many injured, but seeds from deep below the earth had been brought to the surface of the scarred city:

> Over everything – up through the wreckage of the city, in gutters, along the riverbanks, tangled among tiles and tin roofing, climbing on charred tree trunks – was a blanket of fresh, vivid, lush, optimistic green; the verdancy rose even from the foundations of ruined houses. Weeds already hid the ashes, and wild flowers were in bloom among the city's bones. The bomb had not only left the underground organs of the plants intact; it had stimulated them.[19]

The history of eelgrass in the waters off Long Island, New York, demonstrates how a plant can seem a nuisance weed, but when it disappears its previous useful function is revealed. Eelgrass roots in sand and mud and its leaves had once been used to pack seafood. In its dried form it was used for stuffing mattresses and upholstery. However, it could seem unsightly on beaches, cause problems for swimmers and sometimes wrap itself around the propellers of motor-boats. Then suddenly in 1931 eelgrass died out. From the Carolinas to Labrador, in Sweden, Holland and also in English and French water, a waterborne fungus attacked. Migrating wild birds – and the Brent goose in particular, which fed exclusively on eelgrass leaves – were left to starve. Eelgrass had provided nursery grounds for young fish, lobster, clams and mussels.[20] Even their decayed growth had provided a rich sludge on which smaller creatures had fed. Without its root mass the sand had lost its essential role of anchoring the dunes, and fish stocks were down. When the plant gradually began to return in the 1940s and '50s there was a better understanding of the part it had played in biodiversity. Eelgrass was no longer thought of as a nuisance weed.

Marram – or *Ammophila*, from the Greek, meaning sand lover – is a tough, metre-high grass that grows in sand on north Atlantic beaches. Its glossy, tightly curled, spiky leaves protect it from drying out and its deep root system anchors the sand in place. In seventeenth-century Britain it was used in coastal areas for thatching roofs. Gradually it was introduced to colonial settlements in Australasia, the Falklands and also Japan, Argentina and Chile, and in the nineteenth century to America's Pacific coast. Yet despite its successful role in stopping sand erosion, the grass is now considered invasive in all these countries.

This is the history of the weed: an abundant indigenous plant that may come to be valued for some useful property, is thus widely transported, and is then denigrated as an alien interloper.

three

Image and Allegory

After 100 years of enchantment a princess lies sleeping in a castle surrounded by wild woodland, by twisted briars and brambles, but the undergrowth is alive with possibility. In Stella Gibbons's *Cold Comfort Farm*, the spring vegetation suggests a new beginning for Ada Starkadder, weeds announcing sensuality and change: 'The sukebind hung in the hedgerows, its heady scent a promise of the summer and the deeper, darker yearnings growing within her.'[1]

Princess Aurora is waiting to be brought back to life too. When a certain prince approaches, this hitherto impenetrable weedy barrier lets him pass. In Arthur Rackham's silhouette drawings of 1920, the wicked thirteenth fairy is shown living alone in a ruined tower clogged with weeds. In the rough stone walls, nettles and deadly nightshade are growing. Now the great trees make way for the prince, revealing the bodies of former suitors, impaled upon thorns. In one corner, relatively gentle, cow parsley, daisies and cowslips grow. He reaches the princess, kisses her, she wakes and they instantly fall in everlasting love. It is a tale of wild nature, wild weeds responding to human desire. The princess has been both protected and trapped by the wild wood undergrowth:

> Thorns which looked so sharp and cruel became soft as thistledown as soon as he touched them, and the trailing bramble branches did not entangle him but bent aside at his

touch as though they had been stems of grass. The hedge opened before him.[2]

Considering our relationship to the natural world and to plants in particular, weeds, with their sometimes supposed villainy, sometimes grace and beauty, seem to define the ambiguity of our own position. We are part of nature but want to consider ourselves separate, or at the very least a cut above. Our attempts to reproduce and fictionalize weeds in works of literature and visual art may begin to untangle this inherently contradictory puzzle.

The biblical parable of the good seed implies that we too may fall by the wayside, land on stony ground or have our efforts to thrive be choked with thorns (Matthew 13:1–23). In the parable of the

Three farm workers sleep soundly as the Devil sows weeds in the wheat. Engraving by Crispijn de Passe the Elder, 1604.

The brambles allow the prince to pass and rescue the princess Aurora. Illustration for *The Sleeping Beauty* by Arthur Rackham, 1920.

wheat and the tares – thought to be darnel, a type of ryegrass – an enemy plants tare seeds in a newly sown wheat field in the dead of night. Jesus recommends leaving the darnel to grow until harvest time, 'lest while ye gather up the tares, ye root up also the wheat with them' (Matthew 13:24–30). The question remains as to whether some people are essentially like unwanted weed seed, or whether tares may become wheat, and vice versa.

This parallel, of an individual as crop or weed, has caught the imagination of many commentators. The idea of weeds planted among good seed lends itself to a comparison between good and evil in mankind. In French, the phrase *jeter ses premiers faux* is to commit the follies of youth, literally to cast or throw them down like seed upon the ground, as in sowing one's wild oats. In Danish the expression is *Lokkens havre*, Loki's wild oats, as if the mythological trickster god were himself sowing the mischievous tares. St Augustine takes a clear, uncompromising line:

> The darnel weeds are the children of the evil one. The enemy who sowed them is the devil. The harvest is the end of the age, and the reapers are angels. As therefore the weeds are gathered up and burned with fire; so will it be at the end of this age.[3]

Martin Luther, on the other hand, seems to have more time for weeds: 'Although the tares hinder the wheat, yet they make it more beautiful to behold.'[4] He also reminds us that mistakes can easily be made in identifying what plants are truly weeds, implicitly comparing them to hasty decisions about who should and who should not be put to death as a heretic. John Milton, in a similar vein, suggests that it is 'not possible for man to sever the wheat from the tares . . . that must be the Angels' ministry at the end of mortal things'.[5]

These arguments for tolerance, however, still represent weeds as troublesome entities. Luther, an apologist for women's rights in his time, nonetheless elsewhere compares women to weeds: 'Girls begin to talk and to stand on their feet sooner than boys because weeds

St Anne teaching the Virgin Mary to read, framed with rambling wild plants, 1430–40.

always grow up more quickly than good crops.'[6] So, weeds are bad crops, thus women are inferior.

In Chinese, and to a lesser extent Japanese, art one can see examples of an interest in the particularities of different plant species, including weeds, as early as 950–1250 in China. These finely drawn images of the real world gradually became monochrome, with large landscapes as well as much smaller depictions on ceramics, fans and lacquerware. From the Yuan dynasty a silk scroll of 1321, for instance, depicts insects and plants painted in ink and colour on silk. Among the dragonflies alighting on chrysanthemums, there are finely painted bamboo and willow leaves, a dandelion showing flower and bud and, immediately recognizable and most detailed here, that common weed of the European lawn, a plantain, with its oval leaves in a flat rosette, its spikes of tens of tiny bud-like flowers along their length. On a silk album leaf from the Ming dynasty (1368–1644) is a member of the Asteraceae family, similar to the common daisy, but here the weed is a background to two fighting birds, yet still anatomically correct.

Chinese woodblock prints were introduced to Japan as early as the eighth century but Japanese fine woodblock depictions of plants were limited in number until the nineteenth century. A sixteenth-century pair of Japanese paper scrolls, decorated in ink, is attributed to a follower of the artist Sesshu, and includes plants painted with typical restraint, yet accurate in their detail. A wild goose flies over a pond, for example, and in the foreground you can see a lotus seedpod, and make out the seeds within the little circular chambers on the flat top surface. Another, in the background, is drawn typically erect, but the close-up seed head has begun to bend over so that we can see the detail, but also because this is what happens just before the seeds are released.

In the West until as late as the Renaissance, generic plants and flowers in art were largely all that was required, stylized in form and unidentifiable. In medieval books of hours plants are decorative features, their exact type and form immaterial, 'springing up with other *mille fleurs* . . . between the hooves of tapestried unicorns, or at the feet of the Virgin in her *hortus conclusus*', enclosed garden.[7] Albrecht Dürer's watercolour *Das grosse Rasenstück*, the *Great Piece of Turf* of 1503, is said to be the first realistic representation of weeds in Western art, with minutely accurate studies of dandelion, plantain, daisy, speedwell, yarrow and hound's tongue, drawn from observation. There is a collection of Leonardo da Vinci's drawings in the Royal Library, Windsor, showing rearing horses, rocks in all their minute detail, even the uterus of a cow – and among these precise studies are various plants, with notes alongside of their habits. Many were incorporated into paintings and sculptures, though the amount of attention given to the smallest structural detail seems to go beyond what he required for such later use. Leonardo drew marsh marigold and wood anemone,

Leonardo da Vinci, detailed drawing of a marsh marigold with three flowers beside a wood anemone, *c.* 1505–10.

Thomas Gainsborough's study of a foreground, with a bank with thistle and other weeds, 1742–88.

for example, each plant carefully angled to reveal separate specimens, showing the precise form of the flowers and leaves in pen and ink and chalk. He included wood anemone with sun spurge and also brambles in studies for his painting *Leda and the Swan*, of 1505–10. In 1625, before the painting was in all likelihood destroyed, Cassiano dal Pozzo described 'the plantlife . . . rendered with the greatest diligence'. This attempt to depict accurately the world as it is is perhaps a feature of humanism in the West, revealing an interest in the individual dignity of a human being and even of something as lowly as a weed, in all its disparate detail and abnormality.

A hundred years later, in Caravaggio's painting *Still-life with Basket of Fruit*, there is further evidence of a desire for realism in art, showing fruits that have not been idealized, a small apple with wormholes, diseased leaves and signs of imminent decay – nature exposed, warts and all.

By the eighteenth century, it had become normal for artists such as Thomas Gainsborough to include accurate studies from life in their painting of weeds and wild grasses. At the end of the eighteenth century William Kilburn produced a finely detailed print of a dandelion. He is responsible for producing most of the plates for the first volume of William Curtis's *Flora Londinensis*, showing 'plants and descriptions of such plants as grow wild in the environs of London'. It incorporates the new Linnaean binary system of plant names, which Pope Clement XII found distasteful, making too much of plants' sexual nature, he believed. A Japanese *fukusa*, or gift cover, of about the same period, finely embroidered in silk and gilt thread, includes nettles, shepherd's purse, Japanese parsley and cottonweed and chickweed, traditional symbols of longevity and good health, and similarly precise in their lifelike detail.

The Pre-Raphaelites were drawn to weeds as a potent natural symbol. William Holman Hunt's *The Light of the World* (1851–3) has Jesus knocking at a door that is overgrown with weeds and brambles. The weeds are half dead, suggesting a barrier between Jesus and whoever lies within, and said to symbolize 'the idle affection' of the soul.[8] In

A silk *fukusa*, or gift cover, embroidered with bee nettle, shepherd's purse, Japanese parsley, cottonweed and chickweed, traditional symbols of longevity and good health, 1750–1850.

their brittle state, they would be easy enough to push aside were anyone to try and open the door. Hunt further explains their significance: 'The closed door was the obstinately shut mind, the weeds the cumber of daily neglect, the accumulated hindrances of sloth.'[9]

Hunt's *Our English Coasts*, which shows a group of sheep which have strayed perilously close to a cliff edge, was influenced by John Ruskin who, in *Notes on the Construction of Sheepfolds* in 1851, compared Christians to strayed sheep, 'always losing themselves . . . getting perpetually into bramble thickets'.[10] Richard Mabey points out that whatever the religious significance of the weeds might be, the foreground sheep that are caught up in the brambles are at less risk, so in a sense the weeds are protecting them.

With the development of the camera, lifelike studies could be much more easily produced. Karl Blossfeldt (1864–1932) originally took photographs as a teaching aid for his drawing classes at the Berlin College of Art. The architect and designer August Endell describes Blossfeldt's work as 'exquisite curves of blades of grass, the miraculous pitilessness of thistle leaves, and the callow youthfulness of shooting leaf buds'. These close-up photographs of plants taken out of their natural context might be said to reduce them to abstract forms. Blossfeldt focused on the detail. Yet his plants are not perfect forms from a botanical garden or a florist, but gathered from the roadside or railway embankment. In Germany in the early twentieth century there was an upsurge of interest in man's relation to nature and naturism, in open-air swimming and gymnastics, for example, and in an appreciation of the naked human form. Blossfeldt's abstracted images laid plants bare, their symmetries and asymmetries exposed and magnified. It was 'often the plants generally and unjustly denigrated as weeds whose form fascinated him most'.[11]

He was just as interested in weeds that were beginning to wilt and die. Not soft-focus, romantic versions of the natural world these, but sharply caught, vulnerable, rotting, diseased, ageing reality. Working in black-and-white alone, he provided no colour to distract from this cool vision. Compare his images with those of William Bradbury, say, who in 1854 was using actual specimens to create prints, such as his *Nettle*, which one might have thought would retain more of the vibrancy of the original, and certainly be more realistic – yet the effect is strangely less intense and revealing of the intimate details of each plant.

One contemporary writer found that photography was her means of getting up close to the detail of burdock, a stately weed of damp waste places and hedgerows, its lower curling heart-shaped leaves particularly distinctive with their downy grey undersides. Janet Malcolm captures the flowers, round and purple, surrounded by a globular

Two of Karl Blossfeldt's architectural images, of an empty head of knapweed and a silkweed flowerhead.

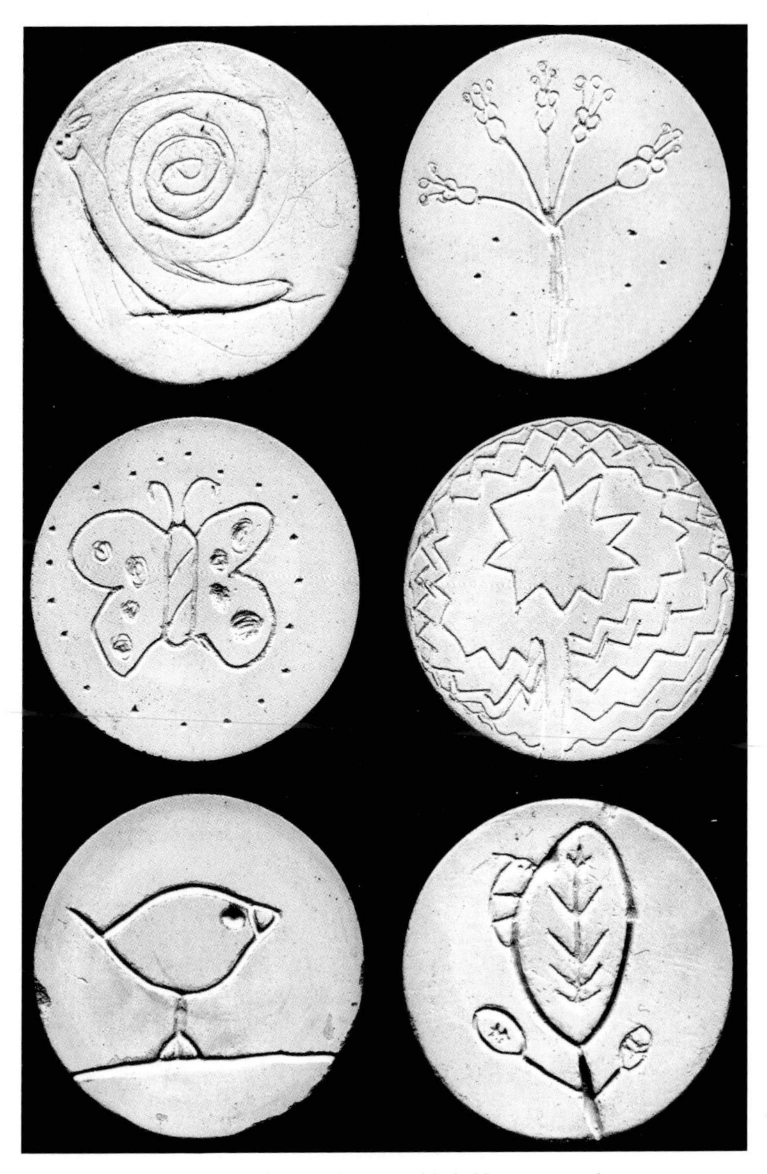

A school project for 7- and 8-year-olds, led by Danuta Solowiej.
Plasters inspired by Karl Blossfeldt's photographs, 2013.

envelope of long, stiff scales with hooked tips, often interwoven with a white down. Their name, *Arctium lappa*, from the Greek *arktos* meaning bear, suggests the rough texture of their burrs that so easily catch onto any passing animal, *lappa* derived from the verb to seize – onto fur or feather, human skirt or trouser leg. Its common name too suggests its method of dissemination. 'Burr' is a contraction of the French *bourre*, from the Latin *burra*, meaning a tuft of wool. Burdock burrs catch easily onto sheep. The way the seeds clung to the underbelly of a dog inspired a Swiss engineer, George de Mestral, with the idea for Velcro fastenings in 1941, perhaps the most poetic fact about that unlovely form of fastening.

The burr also lends itself to the idea of romantic love. Burrs are hard to disengage, and certainly cannot be merely shrugged off, but they are not so hard to remove as love, as in Shakespeare's *As You Like It* (Act 1, Scene 3):

> ROSALIND: How full of briers is this working-day world!
> CELIA: They are but burs, cousin, thrown upon thee in holiday foolery. If we walk not in the trodden paths, our very petticoats will catch them.
> ROSALIND: I could shake them off my coat. These burs are in my heart.

Burdocks are couched in the foreground of paintings by the Dutch painters Jan Wijnants and Jacob van Ruisdael in the seventeenth century, and in the eighteenth in Gainsborough and George Stubbs, for instance.[12] And what should foregrounds give us but context? Ruskin assumed that this was the function and preordained purpose of such foliage, 'the principal business of that plant being clearly to grow leaves wherewith to adorn foregrounds'.[13] A commonplace plant lies between us and a fine eighteenth-century family group in silks and satins, say, or a fine thoroughbred horse, elegantly posed.

In counterpoint here, one might consider the *Bouquet* series of films of the experimental filmmaker Rose Lowder (b. 1941), with her explosive vision of plants in the French countryside. The tables are turned, with a foreground context of goats, cows, farm cats and so on, and the weeds are centre stage and far from passive.

J.M.W. Turner has summer field or hedgerow plants, including cow parsley and dock, in his sketchbook studies of wild plants.[14] His sketch of *Calais Sands* at sunset requires, according to Ruskin, 'the line of old pier-timbers, black with weeds'.[15] Richard Mabey suggests that for his *Study of the Leaves of Burdock*, Turner may have been drawn to the plant's sculptural asymmetries. In Janet Malcolm's short essay that accompanies her prints of burdock, she focuses on the leaves over three summers:

> I prop them in small glass bottles and photograph them head on, as if they were facing me. No two leaves of any plant or tree are exactly alike, of course, but burdock leaves are of conspicuous and almost infinite variety. They are also outstandingly large – more than two feet long in some cases – which makes them extraordinarily good photographic subjects.[16]

Malcolm makes a comparison between her photographs and the revealing celebrity portraits of Richard Avedon, for just as he 'sought out faces on which life had left its mark, so I prefer older, flawed leaves to young, unblemished specimens – leaves to which something has happened'.

The photographer Harry Callahan, in contrast to Blossfeldt, often worked outside in nature, sometimes in colour. Yet since they are close-ups or taken from odd angles then they can seem like abstract patterns. *Weed Against Sky, Detroit* (1948) is an abstraction that can become many things. His photographs remind us of looking at any detail. Our eyes can seem to close like a camera's shutter, with too-bright light distorting and adding strange flashes of colour, for with focused attention any object may begin to refract and distort. We can lose a sense of its relation to other things in the world, and like a single

Close Ties by Patrick Dougherty, 2006. Waste willow woven into 'buildings', in a landscape where willow is seen as a nuisance.

Patrick Dougherty's *Call of the Wild*, 2002, with woven willow jugs, reflected in a building housing contemporary art.

word repeated, it may seem to lose meaning. When Callahan uses a silhouette of his wife's body against light containing images of plant and sky, then the point is more succinctly made, of the blurred boundaries between one category of object in the world and another, and for our purposes here, between plant and so-called weed.

When plants that are widely considered weeds are used in an artwork, it is often a deliberate conceit. When the American artist Patrick Dougherty makes vast woven shapes out of waste willow, for example, he does so in the knowledge that willow is considered a nuisance in the northern American countryside, fast growing and greedy for water. At the same time it symbolizes the natural world. *Close Ties* has willow 'buildings' set in an agricultural landscape where they are noxious weeds, a liability to the mustard crop and the freshwater lake nearby:

> I often use nuisance plant material for building my sculpture. Willows in one area of the country are revered and in another

> place hated. On the east coast of the U.S. I often use maple of one variety or another, a seed, which often arrives first on ditch banks or in abandoned fields.

Yet, given a different setting, he incorporates weeds from that new region, and there is a certain relish to his statement of how he values these otherwise denigrated plants:

> In Hawaii, the hated plant is strawberry guava. This is a very flexible sapling and highly prized if I am working there. Kansas has a grey, rough leaf dogwood, a small bush, which overruns the grasslands and Siberian elm, which chokes out the edges of rivers. Both of these species have found their way into some impressive sculptures at universities, art museums and botanical gardens in mid America.[17]

Dougherty's *Call of the Wild* of 2002 has woven jugs, 18 ft (5.5 m) high, which seem to be pouring water into a pool. Nearby are two walls of an modern office building, the windows and the surface of the water reflecting the textures of the willow and maple. These buildings turn out to be the Museum of Glass in Tacoma, Washington, and in this context of unforgiving architectural modernity the crafted vessels stand out as simple, organic forms.

The artist Jacques Nimki took over a derelict shop in 2009, in a rundown part of Corby in Northamptonshire, a county once at the heart of rural Britain and subsequently of industrial steel production. In the window he grew wild flowers and entitled it *I Want Nature*, as if by some lucky chance the seeds had landed there and something of beauty had grown up in the desolation of economic decay. His drawings are complex networks of intertwined weeds and wild flowers. What was desolate had become fertile meadow. The concept has been compared to Agnes Denes's *Wheatfield – A Confrontation* (1982),[18] where she planted a large field of wheat in Battery Park on Manhattan Island, on a landfill site. She is pictured walking through the golden wheat, a staff in one hand like some bucolic medieval farmhand, with behind her a wall of urban skyscrapers with Wall Street and the World Trade Center in view. Both Denes and Nimki are perhaps asking us to confront romantic ideas about nature. Nimki's weeds are a comment on our relationship to the natural world. Like a theme park again, it has been planned and planted and we are not allowed access. We can only peer through the window. Many of Nimki's works are described by the artist as 'Florilegiums', a gathering of flowers. In the seventeenth century *florilegia* were books where the images were considered more important than the text and often described rare, new species. Later the emphasis was on scientifically useful distinctions, rather than a plant's aesthetic qualities alone. Nimki turns this idea on its head, for his rare plants, in the context of a recession-hit high street, are weeds, and of course even weeds may be beautiful in the concrete city.

These images of weeds, representing fragile nature in the modern metropolis, are reminiscent of John Light's touching book for children, *The Flower* (2006). In a dreary dystopian landscape where 'Brigg lived in a small room in a big city', Brigg comes across a book in the library where he works, marked 'Do Not Read'. Inside the book he finds pictures of things he has never seen before: they are flowers. Later he sees illustrations of these same flowers on a packet of seeds in a shop window and, wondering at their potential, he takes them home and plants them. Like Nimki's empty commercial property in Corby, or the barren building site in Denes's Lower Manhattan, the fragile, colourful plants bring their peculiar magic to the urban.

Michael Landy is an installation artist known for works such as *Break Down* (2001), in which he gathered together all his possessions, large and small, carefully catalogued them and then just as carefully broke them down into their basic materials and buried them at a landfill site. He made a series of life-sized etchings of urban weeds, *Nourishment*, in 2003. Like Blossfeldt's work, they are monochrome and seem to be without context, though as viewers we bring our experience of botanical drawings, and an assumption perhaps that a cool accuracy and detail will be important. In contrast, in Lucian

Detail of Jacques Nimki, *Florilegium*, 2009, acrylic on laminate.

Freud's paintings of London back gardens, a tangle of 'ugliness and neglect',[19] the plants are as emotionally fraught as his human portraits. Landy's drawings seem discreet, and appear to make little demand on us. Perhaps after *Break Down* this is a manner of convalescence, looking at usually disregarded weeds with the same meticulous attention he had given his belongings. The more blatant theatricality of his destruction of earthly goods is set against the inherent fragility of plants. These weeds from the pavements of the East End of London near his home – herb Robert, shepherd's purse, creeping buttercup and thale cress, twelve survivor urban weeds in all – might otherwise have been trampled underfoot or sprayed with municipal herbicide. Landy carefully uprooted and nurtured them, but look carefully and there is evidence of the disease and decay that both Dürer and Caravaggio had captured before him. Landy describes the weeds as 'marvellous, optimistic things . . . They occupy an urban landscape, which is very hostile, and they have to be adaptable and find little bits of soil to prosper.'[20]

Jacques Nimki's sketch, on concrete and peeling paintwork, of his *Florilegium*.

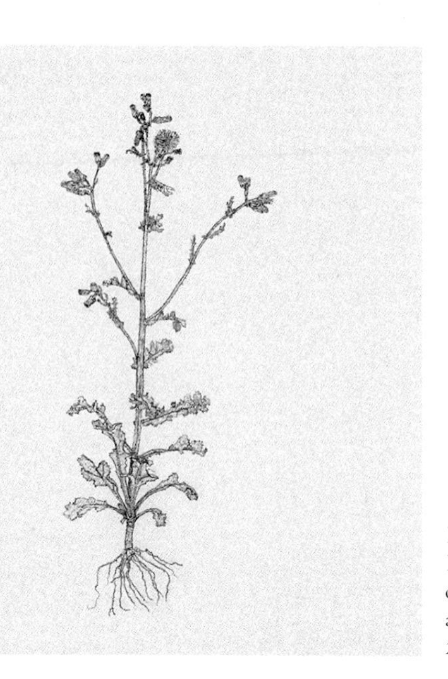

From Michael Landy's *Nourishment*: common groundsel and herb Robert, 2003.

Ovid's *Metamorphoses* describes aconite, created from the spittle from Cerberus' salivating mouth:

> The dog struggled, twisting its head away from the daylight and the shining sun. Mad with rage, it filled the air with its triple barking, and sprinkled the green fields with flecks of white foam. These flecks are thought to have taken root and, finding nourishment in the rich and fertile soil, acquired harmful properties. Since they flourish on hard rock, the country folk call them aconites, rock-flowers.[21]

Aconites produce a strong and fast-acting poison, said to be particularly effective if applied to the genitals, and particularly to the vulva. It is also known as wolf's bane, since it was thought to be powerful enough to kill wolves. In the film of 1931 Van Helsing uses it to protect Mina from Dracula. It was called monkshood in medieval times because of the similarity of the flowers' appearance to a monk's hooded habit. Medea tried to poison Theseus with aconite and in James Joyce's *Ulysses* aconite poisoning is the method by which Rudolph Bloom, the protagonist's father, commits suicide. In a more comic vein the Japanese *kyōgen* play *Bushi* turns on the plot device of dried aconite root; in the *Harry Potter* novels, werewolves rely on a potion brewed from aconite, which they of course refer to as wolf's bane.

The poet Alice Oswald (b. 1966) called weeds the 'flora of the psyche', suggesting that flowers can be 'recognisably ourselves elsewhere', so that on the one hand there is a comic connection being made between toadflax and a 'Ponderous, obstinate / cold-skinned person', but on the other a more serious personification of vegetable nature, with all the cunning qualities peculiar to humanity. In *Hamlet*, the prince's life seems to him like 'an unweeded garden / That grows to seed; things rank and gross in nature'. The ghost warns Hamlet that if he fails to avenge his father, then 'duller shouldst thou be than the fat weed'. Hamlet threatens Gertrude, 'do not spread the compost on the weeds / To make them ranker' and Ophelia sings about the

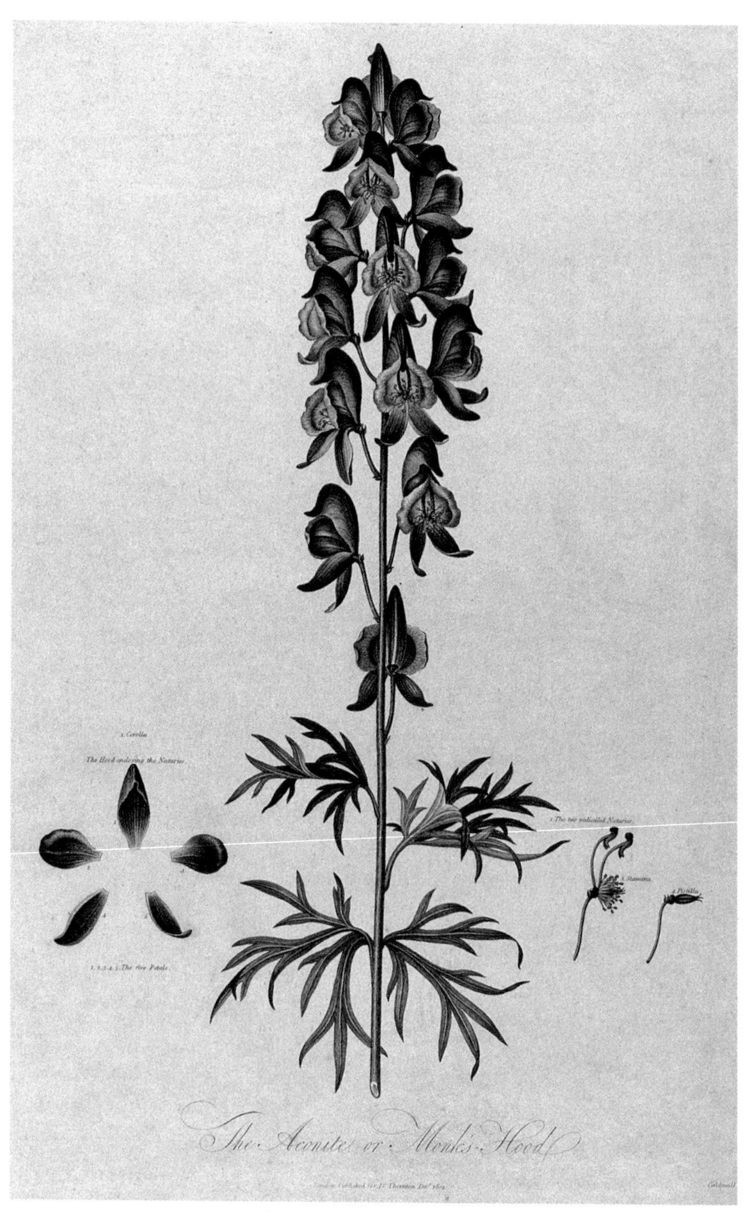

A flowering stem of aconite, with separately labelled floral segments. Engraving by James Caldwell, *c.* 1804, after Peter Henderson.

Richard Redgrave, *Ophelia Weaving Her Garlands*, 1842.
She holds a poppy bud, a symbol of death.

corpse of true love as 'larded with sweet flowers'. In *King Lear* weeds betoken disorder, suggesting a person who is no longer in control of nature, but himself becoming part of its wild vegetation. Lear in his madness hides from the one daughter, Cordelia, who is trying to help him, shrouding himself with weeds and flowers:

> As mad as the vexed sea; singing aloud;
> Crowned with rank fumiter and furrow-weeds,
> With burdocks, hemlock, nettles, cuckoo-flowers,
> Darnel, and all the idle weeds that grow (Act IV, Scene 4)

A Midsummer Night's Dream has love-in-idleness (heartsease or viola) producing a potion that works as an essential plot device to trick Titania into loving Bottom and sow mischief among the young couples. In classical mythology Cupid, the god of desire and erotic love, is said to have accidentally shot his arrow into the plant, thus imbuing it with its magical properties. For the French they are *pensées*, idle thoughts, messing with the reasoned mind. Oberon explains to Puck:

> . . . marked I where the bolt of Cupid fell.
> It fell upon a little western flower,
> Before milk-white, now purple with love's wound,
> And maidens call it love-in-idleness. (Act II, Scene 2)

Oberon considers the best place to render Titania 'full of hateful fantasies' is a weedy, wild flower bed:

> I know a bank where the wild thyme blows,
> Where oxlips and the nodding violet grows,
> Quite overcanopied with luscious woodbine,
> With musk roses, and with eglantine.

Thus the fairy queen lies on wild thyme, primroses, violas, wild honeysuckle and strains of wild roses and is anointed with the magic potion.

Gardens become a political metaphor in *Richard II*. The Queen overhears two labourers as they discuss the state of the country while pretending to criticize only the management of the garden. They complain about having to waste their time weeding flowerbeds, when weeds are being allowed to grow nearby.

In Shakespeare's sonnets weeds are often symbols of rottenness and corruption because they threaten cultivated plants and are said to corrupt the psyche. In Sonnet 69 the object of affection appears beautiful but may be contaminated either by the ground in which they grow and live, soil taking on a sense of what is soiled, or blemished by hidden corruption:

Carlotta Leclercq as Titania in *A Midsummer Night's Dream*, at the Princess's Theatre, London, 1856.

> To thy fair flower add the rank smell of weeds:
> But why thy odour matcheth not thy show,
> The soil is this, that thou dost common grow.

Similarly, in Sonnet 94, a comparison is drawn between the beloved and a festering plant. Weeds are a threat to an agrarian economy, and on aesthetic grounds their foul smell suggests, figuratively speaking, an unpleasant reputation, some inner foul essence hidden by the beautiful outer carapace:

> For sweetest things turn sourest by their deeds;
> Lilies that fester smell far worse than weeds.

Many weeds' common names retain this association, such as with stinking horehound, the devil's stinkpot, stinkhorn, stinking chamomile or stinking hellebore.

When weeds appear in city landscapes in fiction, they are often there to heighten a sense of moral foulness. In Thomas Mann's *Death in Venice* (1912), for example, Gustav von Aschenbach becomes lost in the narrow streets and stumbles on weeds growing between the cobbles and stinking rubbish, symbolizing his own debased dignity. In *A Portrait of the Artist as a Young Man* (1916) by James Joyce, in a vision of his own sinfulness, perhaps, Stephen Dedalus has a vision of a weedy field covered in excrement in which nightmarish goat-like creatures live.

In August Strindberg's *Miss Julie* (1888), weeds help express the servant Jean's sense of humiliation and disgust at his attraction to his mistress:

> I saw a pink dress and a pair of white stockings. You. I lay down under a pile of weeds, *under* – can you imagine that? – under thistles that pricked me and wet dirt that stank. And I looked at you as you walked among the roses.[22]

Thoreau takes a Transcendent view, determined to see his farming project as a means of personal growth, so that weeds, although troublesome for the farmer, are a blessing for hungry birds. He can even describe the sound of his hoe on the ground as music, with beans as the audience: 'My hoe played the *Ranz des Vaches*', a French folk song.[23] Gerard Manley Hopkins sought to discover God in all aspects of nature including weeds, 'a strain of the earth's sweet being in the beginning /

In Eden garden', though towards the end of his life he began to doubt that nature could withstand the assault of the modern world. Weeds became a nostalgic vision of nature unspoiled by man. The alliterations of the second line of 'Spring' (1876) seem to express this sense of longing for the past:

> Nothing is so beautiful as Spring
> When weeds, in wheels, shoot long and lovely and lush.

In 'Remembrances' John Clare captures something of the oxymoronic nature of weeds 'troubling the cornfields with destroying beauty', as symbols of the outcast and mental illness but also of past beauty. Weeds represent what has been lost in the progress of agrarian reform at the end of the eighteenth into the nineteenth century in the West, of all that had been destroyed by man, by the 'blundering plough'.

Weeds are a metaphor for the seizure of the common land in England. Those who were in favour of enclosure saw weeds as a nuisance to be conquered by modern farming. Clare was opposed to changing the old ways of sharing common land in the countryside and saw himself at one with despised wild plants, as in 'To An Insignificant Flower, Obscurely Blooming in a Lonely Wild' (1820), where weeds are 'wild and neglected like me'. Mina Gorji describes Clare's stance, representing weeds as politically and perhaps aesthetically radical:

> For Clare weeds are beautiful, and part of their beauty consists in being outside the system of productivity and profit. They are stubborn reminders of other ways of seeing the world. In one sense perhaps they are a symbol of poetry, art for art's sake, but more than this, they represent an idea of art that is not anthropocentric; they are an acknowledgement of a more enduring order of things.[24]

Weeds flourish in the neglect and litter of an urban street.

The poetry of A. E. Housman, Edward Thomas, Gerard Manley Hopkins and John Betjeman and many more uses the language of weeds as complex allegory. Even *Harry Potter* has the *bubotuber* weed, resembling a large black slug, whose pus can heal acne.

There has long been a fear that plants may have it in them to resist our human needs. Twenty-first-century science warns against the effects of global warming and the heat-trapping properties of cities leading to even more robust strains of weeds.[25] The idea of the superweed has existed for many years in the realms of science fiction, before the reality of plants adapting to herbicides and the GMO debate in Europe. An extensive range of horror books and films evokes plants out of our control, such as John Wyndham's *The Day of the Triffids* (1951), whose monster weeds, unfettered to the ground by their roots, escape and run wild, with a green, venomous sting that can blind and kill, feeding on the flesh of their victims. Other rampant carnivores include the suffocating rytt vines in A. E. van Vogt's *War Against the Rull* (1912) and those that can mimic human speech, as in Scott Smith's *The Ruins* (2006), to achieve their murderous aims. There are plants that invade

Vigorous bindweed, found in garlands in ancient Egyptian tombs, symbolizing rebirth and fertility.

Nettles, sacred to the god Thor, were burned to invoke his protection during a storm.

from other planets, such as H. G. Wells's Martian red weed in *The War of the Worlds* (1897), which is able to rapidly choke waterways, creeping 'like a slimy red animal across the land, covering field, and ditch, and tree, and hedgerow, with living scarlet feelers'.

Folklore ascribes to weeds mythological powers, often based on their physical characteristics, so that for instance the devil's stinkhorn, or stinkpot, is devilish on account of its foul smell. In many cultures nettles and their sting are associated with power. The Norse god Thor is represented by nettles. His thunder can be avoided by the burning of nettles. Hans Christian Andersen picks up on this sense of supernatural power in *The Wild Swans* (1838), in which a princess must fashion shirts from stinging nettles taken from graves to save her eleven brothers, suggesting that the author may have had an idea of the plant's ecological needs, since decomposing bodies produce both nitrogen and phosphate. The princess must pull the nettles up with her bare hands and trample them with bare feet to extract their fibres, in effect bravely suffering the attack of weeds and nature.

> With her delicate hands she groped among the ugly nettles. These stung like fire, burning great blisters on her arms and hands; but she thought she would bear it gladly if she could only release her dear brothers. Then she bruised every nettle with her bare feet and plaited the green flax.[26]

Weeds found in water – freshwater weeds and seaweeds – are sometimes depicted as causing, or at least aiding, drowning. After Billy Budd is hanged in Herman Melville's novella, the 'oozy weeds' draw him down into the depths, an innocent young sailor not resurrected into heaven as in Christian ideology but caught deep down on the ocean bed among the sinuous weedy roots.[27]

Silk-embroidered postcards and cigarette cards sent home in their thousands by troops during the First World War as keepsakes, and often serving as final mementos, often depicted flowers. Far out-numbering roses and border hybrids were wild flowers – the field poppy of course, but also wild pansies/violas, common daisies, thistles and dandelions. These tough little weeds, some of which flourished even in trench conditions, must have seemed more appropriate to the men and perhaps more suitable to those at home.

Weeds never perish

GERMAN PROVERB

One year's seeding, seven years weeding

AMERICAN PROVERB

The corn is not choked by the weeds but
by the negligence of the farmer

CHINESE PROVERB

It is a lazy man who says, 'it is only because
I have no time that my farm is overgrown
with weeds'

NIGERIAN PROVERB

A man of words and not of deeds
is like a garden full of weeds

ITALIAN PROVERB

On fat land grow foulest weeds

ROMANIAN PROVERB

To an optimist every weed is a flower;
to a pessimist every flower is a weed

FINNISH PROVERB

four

Unnatural Selection: The War on Weeds

Gardeners and farmers can become obsessed with weeds and their destruction. Some of the same species of weed have followed me wherever I have lived – in and around London with its heavy clay soil, in Midlands loam, in the rich red soil of Nairobi and the thin volcanic ash of central Honshu. It takes a very small effort to pull up herb Robert, a type of cranesbill and weed cousin of the geranium, with its dainty pink flowers, beautiful pointed seed pods and leaves like ferns that stink of cat when crushed. It self-seeds every year, and if I am not careful fills every available space. And yet . . . it offers little competition, it flowers repeatedly for eight months, and I tell myself that it keeps the soil from eroding. Similarly, annual mercury is everywhere. Wild euphorbia seeds burst like popguns on my windows and the rough leaves of green alkanet dominate my dark back garden, fooling me for years that they were the herb borage, their bright blue flowers the same as those that had once floated in a Crusader's stirrup cup before battle to give him courage. Campanula swathes my flowerpots with lighter blue flower heads. Enchanter's nightshade is delicate in appearance with tiny pink-white flowers, and yet it runs riot underground, for with its tiny white spaghetti of roots it can snap off all too easily, forming new plants, however hard I try to trace it down into the earth. Despite denying it light by continual removal of any above-ground growth, a single juicy dandelion sprouts again and again by my front door.

Green alkanet, a bristly perennial and scourge of many western European gardens, sometimes confused with borage.

Borage, the herb whose bright blue flowers might once have floated in a Crusader's cup to bring him courage in battle.

The many ways that weeds proliferate are their strength, but also provide the key to their undoing. Understanding their means of reproduction means it becomes possible to limit their spread. Whether annual reproducing via its seed, biennial or perennial, the root system of a truly successful weed needs to be adapted to its environment, to give it the best chance of survival. Stolons, sometimes called runners,

Bindweed can reproduce by stolon or rhizome, and its prolific seeds can remain viable for twenty years.

creep horizontally along or just beneath the surface of the soil. The parent and all of its progeny, or *genet*, produce clones which break away to form new plants, as with strawberries or brambles. Many grass weeds propagate by this method, as well as prolific herbs such as mint. If mint is grown for culinary purposes or for its scent, then many of its tougher species need to have their roots restricted in a flowerpot or larger buried vessel with its bottom removed, so that the stolons cannot escape. Hawkweed, with its deep-orange flowers, can reproduce via its wind-dispersed seed and vegetatively through both stolons and shallow rhizomes. It is considered an invasive weed in some areas of North America and Australasia. Rhizomes are swollen underground stems that sometimes grow on the surface, but are usually found deeper down, thus resisting the immediate vagaries of weather conditions. Bindweed, for example, reproduces by stolons, which go on to form rhizomes, and this double defence/reproductive system makes it particularly difficult to extricate.

For many the first thing that springs to mind when weeds are mentioned is their corollary, their gerundive form: weeding, the action weeds demand of us. The price of pretty weeds left to flourish is backbreaking labour. Weeding, prior to mechanization, was intensive work. And even domestic gardening can be spoiled by the never-ending need for weeding. Even if we never actually get round to weeding, the sight of weeds in the border and lawn may make us weary and long for their annihilation. It explains the popularity of concrete patios and decking. We anticipate the aching muscles and broken nails and that sense of a job that can never be completed. The weeds will always return.

Burgundy, in Shakespeare's *Henry V*, draws a direct parallel between a countryside destroyed by weeds and war itself:

> The darnel, hemlock and rank fumitory
> Doth root upon, while that the coulter rusts
> That should deracinate such savagery
> (Act V, Scene 2).

Weeds here suggest the primitive in contrast to civilized farming – a coulter being the iron blade fixed to the front of a ploughshare.

> Wanting the scythe, all uncorrected, rank,
> Conceives of idleness, and nothing teems
> But hateful docks, rough thistles, kecksies, burrs,
> Losing both beauty and utility.

Kecksies are hollow dead shoots, for which Burgundy recommends the cut of the scythe. The state must be on guard to avoid political instability; vigilant husbandry is required or the weeds will take over. This 'idleness', this mother of vice, encourages man to sin when he should be hard at work tilling the soil.

In the Garden Museum in London the tools on display include various adapted walking sticks: a late nineteenth-century one with a pruning saw and weeder, hidden beneath a wooden sheath when not in use; a 'spudder' from about 1920, incorporating a spade with a

Jean-François Millet's *Man with a Hoe*, his painting of a man at the age-old, backbreaking task of hoeing thistles, 1860–62.

narrow blade for weeds and roots; a 1930s 'slasher', for cutting away grasses; and, most ingenious, a stick with an integral shaft that could reach into the soil to remove deeper roots, like an instrument of clandestine Cold War assassination. The ricin poison fired from an adapted umbrella into the Bulgarian dissident Georgi Markov's thigh in 1978 was itself derived from the castor oil weed.

The modern-day garden shed is likely to house at the very least a spade and fork. With strong blade and dagger tines to winkle their way between roots, the two are capable of adapting to most gardening jobs. A standard rake can be used to loosen early weed seedlings. However, the most basic specialized weeding tool is the hoe. The Dutch hoe, shaped somewhat like a Greek Ω, is able to get more easily between plants to chip away at young weeds and, with a blade across the bottom, nimble enough to sever deeper roots. Its swan-necked version is better suited to a chopping action, allowing a deeper cut. A heavier, tougher cousin, the draw hoe, can hack out still more difficult weeds. The larger spring-tined rake is used to scratch up leaves from the lawn, but is also useful on unplanted seedbeds, particularly if there is a new show of weed seedlings that need to have their roots exposed, to shrivel in the sun. A medieval-style wooden-toothed rake is still the best way to tear out long grass and nettles, 'haymaking' as you go, like a bent-backed weeder straight out of a Bruegel painting. That said, if you have a small enough garden and enjoy the process of weeding at times, then getting down on your hands and knees with an old kitchen knife does the job too.

In India a curved hand blade known as a *kirpi* is an all-purpose tool in field and garden. Due to the recent unpredictability of the rainy season, with heavy downpours often arriving when least expected, either immediately before planting or when seeds may just be ripening, many small-scale farmers in Madhya Pradesh – in the centre of the country and the second-largest state – are turning to old ways, planting mixed seed fields.[1] This *utera* system might combine rice with pulses and millet in order to stand at least some chance of achieving a partly usable food crop. It is a practical small-scale solution but relies on intensive hand-weeding, with different plants needing to be individually recognized, weed distinguished from crop. Field workers in Africa seem more often to favour the machete or the medium-handled hoe with its flat blade at 45 degrees to its handle. A similar, but long-handled, version is favoured in China, and an example was found without its handle but with an obsidian blade still perfectly intact, made by the Romans over 2,000 years ago.

In developing countries today the bulk of those who carry out this labour are women or young boys, rather than people of all ages, including men, as was the case in John Clare's day in England at the beginning of the Agricultural Revolution, and indeed across much of western Europe. Jonathan Bate describes Clare working alongside 'the old women of the village [who] assisted with the weeding and helped the time pass with their tales of "Giants, Hobgobblins and fairies"'.[2]

Cave paintings in China from 7,000 years ago show men tilling the land with a curved stick – perhaps an animal's shoulder blade or

Weeding in the rice fields, Japan, photographer Shinichi Suzuki, *c.* 1873–83, hand-coloured albumen silver print.

possibly a deer antler. In Shanghai Museum there are three ceramic figures from the Tang dynasty holding what look like modern hoes. Wooden models of men with hoes were found in Egyptian tombs from the Middle Kingdom (2040–1750 BCE), often with their tools held crossed over the body to show they were ready and waiting to help the deceased in the afterlife, should their supply of food begin to dwindle. Wooden hoes have been discovered in Thebes, dating from 1550–1070 BCE. According to the British Museum, the ground would have been ploughed with a wooden plough, and the hoe used when the ground was either too hard or too weedy. Stone blades, their wooden handles long since rotted away, were found in Mesopotamia, dated to 5000 BCE; iron hoes found in Zambia have been carbon-dated to 1200 BCE. In old farmers' almanacs from the American Revolutionary period, rewards were posted for the return of stolen hoes, so great was their market value.

The Iron Age had made specially designed metal tools possible. The European medieval weed hook, for instance, consisted of a pair

An African boy with polio weeds stony ground using a long-handled hoe, *c.* 1950.

of long-handled prongs, one forked and the other tipped with a metal hook. The problem is that all this armoury of weeding tools cannot effectively remove perennial deep or spreading root systems such as coltsfoot, and those weeds whose tiny sections of root or runner can so easily each become a new plant if left behind. In fact the process actually promotes strong new growth, as is the case of nettles and bindweed. Weeding tools cannot suppress the germination of a windborne seed.

Virgil's *Georgics* of 29 BCE describes man's struggle against a hostile natural world where vines become entangled 'unless you continually attack weeds with your hoe', or 'useless thistles flourish in the fields: the harvest is lost and the savage growth springs up, goose-grass and star-thistles, and, amongst the bright corn, wretched darnel and barren oats proliferate'.[3] Alice Sturm, working on an arable farm in Pennsylvania in 2012, described this ancient drudgery as 'one of the most time-consuming and least intellectually challenging tasks'.[4] But she also suggests it allows her welcome time for contemplation. For many the ongoing task of weeding our gardens is one that brings a rare sense of peace. When we separate out each new growth, fair from foul, we may feel we are benefitting the plants we want to encourage, making way for their new growth. Yet the task is a simple one and our thoughts may have a tendency to stray. Focusing on a problem can be unproductive and so the simple, methodical task of weeding may allow ideas that might otherwise escape us.

Manual weeding involves close contact with plants that can result in respiratory and other health problems, such as allergic skin reactions like nettle rash. At the extreme, poison ivy, giant hogweed, sumac and poison oak cause severe irritation and blistering; the poisonous deadly nightshade can result in pustules if you are weeding near a hedgerow and happen to rub against any part of the plant; ragweed, common euphorbia and many sappy wild plants can cause impaired vision, painful rashes and even permanent scarring.

Under the Dock Leaves by Richard Doyle, 1878: elves and fairies dance in a glade, their insubstantiality and scale set against the large, broad-leaved plants.

Weeding is often considered women's work. Here forestry seedlings are being weeded in Tucker County, West Virginia, 1940. B. W. Muir commented at the time, 'Women are used for weeding here because of the dexterity of their hands. They are paid 40 cents an hour.'

Pehr Kalm, the Finnish naturalist and one of those sent out by Carl Linnaeus to bring back seeds, specimens and information to Sweden from all over the world, visited England in 1748. Kalm documents the use of grass cuttings as a hot mulch to warm the soil and when rotted down being used as a nutrient and as weed barrier. In Chelsea he notes how market gardeners gathered together their weeds, and 'laid them in large heaps to use for the same purpose'.[5] On the other hand, he describes fruit tree production, mentioning that their yield was thought to suffer from the weeds at their roots, and explains why peas were sown in rows: 'Weeds which smother and draw food from the peas could then more easily be cleared away with a hoe . . . the peas grow all the better for the soil being so friable and loose.'[6]

The plough was traditionally used to destroy weeds before sowing because

> it kills the Weeds by turning up the Roots to the Sun and air, and kills not only the Weeds that grow with the last Corn; but wild Oats, Darnel, and other Weeds, that sow themselves, and that as soon as they begin to peep out of the Ground, so that they have no time to suck out any of the Heart of the Land.[7]

Jethro Tull invented not only the seed drill in 1701, but in 1703 the horse-drawn hoe, which more efficiently achieved the laying back process that Kalm had observed, leaving the weeds' roots on the surface to shrivel and die, though deeper-rooted perennials would quickly repopulate without additional hand weeding. Tull likened his husbandry ideas to creating a vineyard-like culture. Wine growers face a battery of weeds that enjoy the same conditions as their vines. In mid-twentieth-century Austria, the vintner Lenz Moser found some weeds hostile but others favourable to grape growth and flavour, and his company today attempts to use environmentally less harmful weed control in order to maintain these helpful weeds' influence on their crop. In the Californian vineyards many virulent weeds have gradually become resistant to herbicide control – hairy fleabane, horseweed or fleabane, Italian ryegrass, rigid ryegrass and jungle rice, to name but a few. Some weeds are in the process of becoming resistant, like barnyard grass, Johnson grass and the annual tall willowherb. As with a build-up of resistance to antibiotics in humans, over-use seems to be the culprit, with farmers applying chemical herbicides two or more times a year. The danger lies in such 'superweeds' becoming almost impossible to kill. The weed scientist Brad Hanson, for example, recommends rotating different applications of the most commonly used glyphosate herbicides with glufosinates and also using other, older methods, such as carefully timed weeding after rainfall, when weeds may suddenly flourish and can more easily be easily dislodged before they become established.[8]

Aquatic weeds present their own problems for fish farmers and also for the environment at large. Water weeds such as *Chara contraria* and *Potamogeton pectinatus* in southern Argentina and the lower valley of the Colorado River can block pump intakes in hydroelectric complexes, or obstruct irrigation and drainage channels, which may lead to flooding and salt water draining onto cultivated land.[9] Freely floating mats of weed can become a navigational hazard, such as the reputed world's worst aquatic weed, the beautiful purple-flowering water hyacinth, a southern American species that has wrought havoc in Florida. It is claimed that a healthy acre of weed weighs up to 200 tons, jamming waterways and lakes with its fleshy matter. Dense aquatic weed may harbour snails that in turn harbour parasitic flatworms, or blood flukes, which carry a chronic parasitic disease in the tropics that in its devastating effects is second only to malaria.[10]

A study carried out in the 1970s on rice paddy fields in Japan found that in a 1-hectare field, whereas traditional hand weeding might take 500 hours, treatment with a herbicide would reduce this to 90 hours, and a naturally occuring tadpole shrimp could reduce this still further to only twenty hours.[11] This minuscule creature was capable of a host of tasks: uprooting small weed seedlings; inhibiting weeds from growing by agitating the soil until the water became

African workers weeding a rice plantation, Cape Fear River, North Carolina. Wood engraving, 1866.

muddy; inhibiting the photosynthesis of small seedlings; eating the young buds and seedlings and even the fungus growing on rice stems. This method works only on transplanted rice fields, as the shrimp do not discriminate against young rice seedlings.

Age-old approaches to weed control are often drawn from folklore, where weeds are depicted as malefactors, to be controlled with powerful magic. Wild plants were often, as has been mentioned, associated with the Devil. Nettle hairs, for instance, intimated the tines of the Devil's pitchfork and fungi suggested the presence of Puck and his fairies in Ireland. Blackberries should never be picked after the end of September, since the Devil is said to pee on them in early October; it is certainly true that the cold weather causes them to lose their flavour, and they become watery and gritty. Hawthorn or

May blossom is believed to be unlucky in the house; it also causes asthmatic attacks.

Some apparently New Age methods of control hark back to ancient knowledge concerning plants' interactions. Digging or planting seed after dark, for example, may appear ritualistic, but turning the ground in daylight may well allow buried weed seeds just enough light to germinate.

The biologist Daniel Chamowitz asks us to consider plants as we do animals, with parallel senses of sight, smell, hearing and touch, and even memory. Dodder is a parasitic weed of the morning glory family. The nature writer Amy Stewart describes a field covered in dodder as having the appearance of being under attack from Silly String.[12] After a seed of it has germinated in the ground, dodder sends up delicate spiralling shoots that wind their way about another plant, injecting minute fungal structures into their stems – of wheat, clover, alfalfa, flax, hops or beans perhaps. The dodder plant's roots then fall away, and it becomes wholly dependent for sustenance upon its host.

Marigolds attract pollinators to an allotment.

Some weed plants can be inhibited by touch, known as *thigomorphogenesis*, such that, for example, 'simply stroking [cocklebur] leaves three times a day completely changes its physical development – even its genetic make-up'.[13] Cocklebur is an annual agricultural weed in North America that spreads its seed by hooked airborne burrs, and is severely poisonous to livestock. How such a method could be used other than in small-scale circumstances is difficult to imagine, but it might be possible to develop the technology to stroke cocklebur before it sets seed in the field.

Perennial weeds like dandelion can be controlled by laying back: letting weeds grow until just before they flower, then cutting and laying the plants over the roots below to stifle new growth, much as Kalm had observed. In the garden or allotment dousing with a vinegar solution can gradually inhibit difficult weeds such as ivy or brambles, as the acid dries and shrivels the leaves. Similarly, vegetable oil can suffocate weeds without being a risk to plants nearby. On the smallest of scales, weeds can be held back on a path by dashing them with leftover boiling water from a kettle or saucepan. Overplant your beds and there will be little opportunity for many weeds to grow up, though this is contrary to what some have described as a traditional working-class aesthetic, where plants should be neatly bedded out, with bare, weed-free soil left in between.[14] Alternatively, in meadowland, or even in the domestic garden, chickens will peck away at succulent seedlings, thereby improving the tilth of the soil, and simultaneously fertilizing as they go, though, like the tadpole shrimp, they will not discriminate between weed and valued plant.

Mulches keep light from weeds and come in various forms. Thick black polythene rolled out over beds, with vegetable crops grown through cut fissures, is long-lasting but has the disadvantage of water run-off drying out the soil beneath. Paper and modern permeable textiles, such as bitumen-based felts, can solve this problem. On the allotment, old carpet and cardboard works well but can look unsightly. Organic mulches, such as leaf compost, manure, chicken litter and woodchips, may not be as effective, but have the advantage of both feeding the soil and not

Horsetail resists many herbicides.

needing to be thrown away once they deteriorate. On clay-bound soil, grit and gravel mulches keep off weeds and gradually work down into the soil, improving drainage. 'Floating' mulches of transparent plastic may not kill the weeds surrounding a crop, and may even en-courage their growth, but it does make it easier to weed them out before the crop itself matures.

Alice Sturm makes the point that even on her farm, it remains context that defines weeds. In carrot or kale beds, 'thistles and burdock and dandelions and Pennsylvania smartweed and chickweed and pig-weed' are troublesome weeds, but even a few yards distant from crop beds and they become a useful resource, perhaps providing blossoms

A straw mulch deters weeds and slugs and improves the soil.

to pollinate the crops, helping to avoid soil erosion or becoming 'simply a beautiful thing'. She describes hand weeding between the rows as connecting her in an unbroken line with the farming ways of ancient Mesopotamia; the hoe, being faster and more efficient, seeming to forge a link with the Iron Age. When she uses a flame weeder, perhaps it is the post-Industrial Revolution period she recalls, attacking weeds that have not yet broken ground. The other mechanical methods her farm uses include 'a tractor implement with wheels of wire that tear up tiny just germinated weeds from the soil, between the rows [and] the shovel cultivators, which dig a furrow of raw earth between crop beds', conjuring a scene as visceral as tooled up Roman gladiators marching into battle.

Richard Mabey suggests that Dorothy Hartley's description of early agricultural weeding in *The Land of England* (1979) would have been pieced together from paintings and her own observations and intuitions, as she imagines the age-old process. The weeder

> uses two sticks: with the first, hooked stick he plucks the weeds out from under the corn stalks, and with the second, forked stick pins the weed's head down under the fork. The weeder then steps one pace forward, placing his foot on the head of the weed, and, with this forward movement, swings the hooked stick round behind him, lifting the root of the weed high out of the ground, before dropping it in line. In this way each pulled-up weed is shaken clear of the soil, and laid with its root over the buried head of the previous weed. Thus, as the weeder goes along the line of the furrows he lays a mulch of decaying weeds alongside the roots of the corn, and forms a line between the rows at least as wide as his foot. Weeding employed a definite rhythm, and the feet of the weeder formed the lines on which much of the reaper's work depended.[15]

It was Hartley who edited Thomas Tusser's 1557 *Five Hundred Points of Good Husbandry* in 1931. Tusser was one of the first agricultural improvers to advocate adjusting seeding rates to discourage weed growth, a method of control that is known as 'cultural' by weed scientists. He wrote in rhyme, which Walter Scott suggested was in order to help illiterate gardeners to remember his advice. His verses for March include:

> Sow barley and oats good and thick, do not spare;
> Give land leave her seed or her weed for to bear.

Just as they do today in rural Madhya Pradesh in central India, European farmers in the eighteenth century hand weeded with hoes, weed hooks and thistle drawers. They used animal-drawn cultivation implements and foraging animals to control weeds, but in both cases they also needed to adjust the culture of weeds. William Marshall's

Rural Economy of the West of England of 1796 remarks on wheat being sown often two months later than in other regions, because 'early sown crops are liable to weeds'.[16]

In pre-industrial societies, cultivated crops have always needed to be hand weeded and, even when ploughs were available, it created an economy reliant on mass short-term, seasonal employment. Weeding came to define the business of farming. The more efficient and thorough the weeding process became, the more danger there was, arguably, of damage to soil structure and the removal of sustaining nutrients, which Sturm vividly terms 'a repeated tearing off of the scab'. A traditional method of compensating for this is to leave fields to themselves for a season, rather as allotment holders might divide their plot into four parts, rotating crops with different needs and then allowing one part to lie fallow. This section would then be planted up with nitrogen-rich weed crops, such as clover or wild lupin, or sometimes mulched or covered with some form of weed-suppressing mat. A comparable fallowing process is also used in livestock meadow management, allowing the grass and other plant life to recover from grazing.

Bare summer fallowing is a traditional method that lost a whole season's output, and was practised in Britain, according to William Marshall, since medieval times. Clinton Evans draws on its history to elucidate current northern American practice, bare summer fallowing having been adopted as the most important method of weed control in Ontario and the prairie West. In the sixteenth and seventeenth centuries a field would be left fallow by English farmers every second or third year, during which time it might be ploughed two or three times, depending on the degree of weed infestation. Eighteenth-century agricultural improvers suggested that this pattern might not be sufficient, and that for some perennial weeds fallowing might have to be undertaken for two continuous years before the stranglehold of weeds such as wild oats could be loosened.

The industrialization of agriculture meant, potentially, turning the countryside into a production line, with the new herbicides thought to be capable of wiping out weeds for good and making such fallowing obsolete. While caustic chemicals work by burning leaf or root, systemic herbicides enter the core of the plant. However, these chemicals are often less efficient at soaking into the foliage initially, needing detergents to help break the surface tension on leaves, and then they may not travel far enough into the root system to kill a weed off entirely. Farmers and gardeners may then increase the dose, building up a chemical residue which risks damaging biological diversity and thus the crops they are seeking to protect. Selective weed killers, such as sulphate of ammonia, target broad-leaved weeds in lawns and in pasture and small-grain cereal crops. Farms and outside public space have come to rely on both selective drenching and carpet bombing of pavements with hormone weed killers.

As we have seen, weeds are not that easy to control. As Sturm points out, it is near impossible to obliterate weeds with the application of a chemical:

A weed sprayer used on an urban railway in Yakima County, Washington.

> This . . . requires special crops built to withstand these chemicals, which, naturally, results in special weeds which can resist as well . . . it changes . . . the spirit of weeding, where we carve out our little plot from the wilderness; it changes the game into a war. When modern chemical farms spray broad spectrum herbicides, they do so much damage, to all plants, that they are forced to find chemical solutions to problems that used to be solved by nature itself. Using mass amounts of chemical fertilizer, much richer in nutrients than normal old compost or composted manure, to restore the nutrients leached from the soil by harsh growing of monocultures in herbicide-sterilized land, is an example of the slippery chemical slope.[17]

It is hardly surprising that since weeds are weeds partly because of their ability to survive, then those for whom a successful herbicide has been found may find a way round the problem. The common groundsel, apparently vanquished by atrazine and simazine since the late 1950s, began to develop resistance, the first confirmed example of herbicide resistance. In Australia ryegrass has become immune to at least nine different herbicides. Stephen Powles of the University of Western Australia in Perth suggests that 'herbicide resistance is a fantastic example of evolution in response to human-induced selection pressure',[18] with plants mutating in leaf shape and waxiness, and even changing their molecular structure.

Some plants inhibit the growth of others by themselves producing natural herbicides. These 'allelopathic' compounds are nothing new. The dodder plant found near a crop can be distracted away by tomato plants if they are grown nearby. Both tomato and wheat contain the chemical beta-myrcene, to which the dodder is attracted. However, the tomato plant also contains two other chemicals, and their combined bouquet is more attractive to dodder than beta-myrcene alone. Wheat also contains a chemical that repels dodder.[19] Theophrastus mentions weeds being inhibited by chickpea crops around 300 BCE; Pliny the Elder in 1 CE has corn chickpea, barley and bitter vetch scorching corn.[20]

A dandelion appears to die, but the chemical spray has not reached its deep taproots.

Weed scientists are investigating ways of using allelopathics to favour crops over weeds. Since weeds often gain ground by growing up before the crops have taken hold, one method is to sow a winter cover crop capable of inhibiting the main season's crop weeds. They can also be used as a mulch during the cropping season, or the allelopathic compounds might be extracted and applied as a herbicide. Furthermore, synthetic herbicides have been derived from these allelopathic species.

Modern farming methods can inadvertently aid weed prosperity. Growth hormones fed to cattle, say, can, when excreted, stimulate weed growth on the dung heap or spread weed seedlings when manure is later laid on crop fields. The flame weeder, and the basket weeder attachment on tractors, can of course weed far faster than by hand, but with speed comes less discrimination, and unless they are scrupulously cleaned they can again spread weed seeds. Not that hand weeding always indicates discrimination. Peter Bowden points out that the task is at once both tedious and labour-intensive, and attitudes can often be as slapdash in effect as hand broadcast sowing, trampling crops and using tools carelessly.[21] I recall my grandfather's exasperation when a bob-a-job Boy Scout had made a tidy job of weeding his raspberry canes by merely tugging off the top growth and leaving all the roots intact.

Some terrain makes it difficult to launch an attack on invasive weeds without the sort of blanket bombing that can damage precious local flora and fauna. The remote Hawaiian island of Kauai is rich in biodiversity. It is one of the wettest places on the earth, with over 220 endemic Hawaiian plant species, 92 of which grow only on that island. But it is a land of serrated cliffs and sudden deep valleys. On the forested Wainiha plateau surrounding Mount Waialeale there is 'an exquisitely layered suite of plants, from delicate ferns to moss-covered trees, catch[ing] the water like a giant sponge'.[22] Native species are threatened not only by ravaging wild pigs, but in the turfed-up, eroded soil they leave behind them, strawberry guava, kahili ginger and the Australian tree fern gobble up the suddenly available clearings.

The fern was introduced to Kauai as an ornamental plant 50 years ago, but today its windborne spores have spread out over the archipelago. On Kauai in 2004, Trae Menard of Nature Conservancy Hawaii attempted to destroy them using paint-guns, nicknamed stingers, armed with pellet-sized bursts of herbicide, pointed from helicopters at the ferns like enemy insurgents below. Today technology allows greater accuracy, and a device slung beneath a helicopter can fire small amounts of imazapyr, a short-lived herbicide that is harmless to animals: 'We treated over 4,000 tree ferns in a 5,000-acre area over a three-year period and we only used 11 gallons of herbicide.'[23]

Keeping this weed growth in hand requires a regular monitoring of their spread. Nowadays reconnaissance missions can record a highly detailed photographic mosaic of the island, meaning that even an individual pinna, or leaflet, on a fern frond can be precisely located. Yet despite all this effort, the tree fern continues to recover and gain ground, so that constant vigilance has to be maintained.

Herbicides can kill a broad spectrum of plants, but create a greatly increased need for fertilizers to replenish the soil. Modern robotics minimizes human labour and increasingly has the potential to be selective in weed control, so that beneficial or harmless weeds that contribute to biodiversity can, at least in theory, be avoided. Robotics are increasingly used to tend crops and are particularly suited to fruit and vine growing, with seasonal farm labour becoming less sought-after employment in the developed world. Activities like crop weeding require intensive bursts of work over short periods of time, and fail to provide reliable, long-term income.

Like military drones, unmanned and relentless, robots fitted with herbicide feeds, flame guns and lasers take over the work of seasonal labourers. Smaller-scale machines cause less damage than super-sized ploughs and combine-harvesters on grain crops. Yet as far as any truce with permitted weeds, it seems there will be little change, with autonomous 'agribots' blasting all with chemicals.

At Osnabrück University of Applied Sciences, Arno Ruckelshausen is developing 'BoniRob', a field robot capable of applying minute amounts of pesticide directly to the leaves of weeds.[24] The science journalist Duncan Graham-Rowe mentions a Danish robot that can map the exact position of weeds growing among crops, aiming to reduce the amount of herbicide used by as much as 70 per cent.[25] It follows that such a selective method, like the helicopter applications on Kauai, which can exclude native species, avoids the unnecessary blanket drenching of crops. Similarly, the robot can 'see', in that it is fitted with a camera able to recognize the shapes of weed leaves. Tiny colour variations alert the robot to the presence of weeds among the crops. Trials of the system, led by Anders la Cour-Harbo at Aalborg University, use cameras 'tuned to pick up parts of the light spectrum that correspond to the reflective signatures of the weeds and crops it is looking for – for example thistle sticks out because it absorbs yellow light more than surrounding beet plants'.[26]

Like face-recognition programs, it can identify parameters such as the precise length, width and symmetry of a distinctive plant part. Svend Christensen at the Danish Institute of Agricultural Sciences at Tjele argues that whereas broad-leaved weeds are relatively easy to

An abandoned barn overwhelmed by Japanese kudzu vine
in rural America.

recognize, grassy weeds are more difficult to differentiate and therefore to log accurately into such computerized systems. With some weed species, which depend on light to break dormancy and germinate, often dependent merely on the light received when the soil is turned for planting, there can be an advantage in using these robots to plant crops, somewhat romantically, on moonless nights. Even the tractor headlights must be turned off, reminding one of guerrilla warfare, for even 'a brief flash of light before the seed is buried again is enough' to get a weed up and growing.[27]

Since the late 1990s there has been a steady rise in the genetic modification of crops, to resist disease, and develop both drought-resistant plants and also herbicide resistance, so that chemicals intended for weeds do not damage crops. There are arguments that careful GM technology could in fact benefit all wildlife, though the temptation to eradicate all plants considered weeds in one context remains. Some weeds manage to fend off the attack by cross-breeding with GM crops and so taking on some herbicide resistance themselves. Wild sunflowers in the United States, when crossed with crop sunflowers resistant to seed-nibbling moth larvae, were found to increase their seed yield by as much as 50 per cent, for example.[28] One solution posited would be to make a GM plant infertile, using the controversial 'terminator technology', so that no cross-pollination with weed cousins was possible, though this might of course threaten the long-term future of these crops. Using a form of contraceptive spray on weeds, such as on wild radish, which infests barley, wheat and rape, is another line of attack.

Another example borrows from developments in cancer research. Ryegrass and black-grass, in Australia and across Europe, carry a gene that is able to resist many powerful herbicides. Robert Edwards of York University compares their immunity to the way some animal and human tumours can withstand cancer drugs. When he transferred the same gene into thale cress, that too became immune. But his solution draws further on human cancer treatment, for after applying a drug used on patients to attack tumour immunity, the cress also became vulnerable again.[29] Although the drug is too toxic, so far, to be used in the field, the idea of treating such superweeds as a disease offers potential ongoing means of plant control.

This comparison between weeds and life-threatening disease evokes the sort of excessive response to plants such as Japanese kudzu, the 'mile-a-minute vine' or the 'vine that ate the American south'. A native of Japan and used for millennia as a constituent of herbal medicine in China, it was first introduced to America at the Centennial Exposition in Philadelphia in 1876, and farmers were encouraged to plant it as a hardy, fast-growing creeper to help with soil erosion and also as a source of fodder. It can grow up to 1 foot a day, and is spreading at a rate of 150,000 acres per year. Recently Japanese knotweed has become one the most feared invasive weeds in Europe, partly because it is near impossible to extirpate; its growth can threaten housing equity and eventually make properties unmortgageable. Its East Asian derivation can seem to add to its image as ruthless and alarmingly foreign. Both kudzu and knotweed are like cancer in that they can appear to come from nowhere and no cure seems reliable.

The ways in which weeds reproduce are crucial to their success, as we have seen. Sexual reproduction can be costly, using up essential resources, whereas when a plant is able to clone itself then vital energy can be conserved. Retaining the possibility of sexual reproduction nonetheless allows for the new genetic combinations that can sustain its existence, producing hybrids that can deal with whatever new challenges it meets. Ability to spread out over all available habitats is also crucial. What is small and round, has four limbs and can jump 200 times its own height? The answer is the microscopic spore of the mare's tail, which has the ability, if subject to sudden dry conditions, to disperse with tremendously efficient alacrity, and as Philippe Marmottant of Joseph Fourier University in Grenoble explains: 'If one jump doesn't do the trick, they can do it again.'[30] And again.

In wild and pastoral land, even after the successful removal of invasive weeds, it can take years for indigenous plants to become re-established. It is not only herbicide residues that can inhibit their growth. If weeds bear tough enough tubers, even when they have been killed off, the dead, fibrous mass can still inhibit the roots of seeds planted in their stead.

five
Useful Weeds

Weeds can be of considerable use to us. And when they are deemed useful any clear distinction between herbs and weeds disappears, a useless, otherwise denigrated plant made valuable in our eyes, its status rehabilitated. It is curious to see how everyday complaints from thousands of years ago tally with our own petty physical concerns regarding spots and rashes, coughs and sneezes, indigestion, constipation, diarrhoea and sexual disorders – all calling on the devices of plants to cure and ease. As with modern drugs, the dosage used to cure might easily become poison if misjudged. Rather as warfarin was first developed as a rat poison but in the correct amount has proven effective against thrombosis, so, despite the miseries caused by poison ivy, as early as the seventeenth century it was said to cure a young Frenchman suffering from viral skin disease and to ease the symptoms of arthritis.

Chinese medicine in Shennong's *Bencao Jing* (Materia Medica) prescribes many plants and herbs such as kudzu, cannabis, rue, motherwort, spurge and mint, either bound with honey in pill form or as decoctions, often in combinations with human, animal and mineral extractions, to balance out the two principles of yin and yang. Horny goat weed leaves, for instance, are helpful for joint pain, blood disorders and viral infections and are used to treat the symptoms of HIV/AIDS, and to correct erectile dysfunction in men and menopausal symptoms in women.

Ancient knowledge about poultices, infusions and decoctions was the beginning of what we now know as medicine. Bramble root

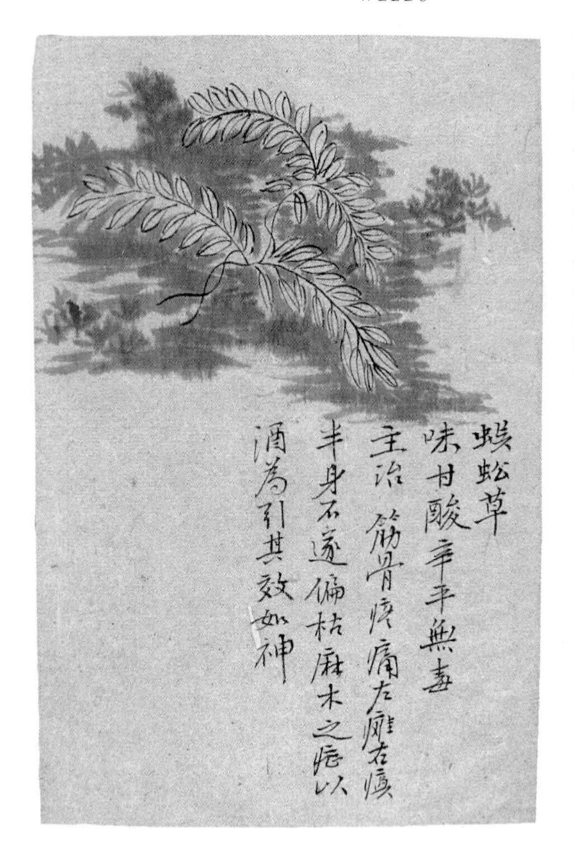

Painting of ciliate desert-grass in *Diannan Bencao Tushuo* (The Illustrated Yunnan Pharmacopoeia), compiled by Lan Mao, 14th–15th century, a record of plants used for medicinal purposes in Yunnan during the Ming period.

can be taken as an effective cure for diarrhoea, plantain juice for haemorrhoids; wild onion, garlic and chickweed serve as an antiseptic. Comfrey, betony and common daisies were ancient wound herbs, the latter known as woundwort. It is said that the great Celtic warrior Cuchulain always carried meadowsweet to ease his fierce temper – and indeed at the end of the nineteenth century meadowsweet and willow bark were used by the German pharmaceutical firm Bayer AG in the creation of aspirin, so called after meadowsweet's earlier botanical name of *Spiraea ulmaria*.

The *Materia Medica* of Dioscorides, the first written herbal guide, draws on Egyptian and Greek learning, suggesting various weed-based cures and palliatives. Anglo-Saxon herbals talk of 'elf-shot' fired at us by supernatural beings and plant potions offered to guard against such vicious flying venom. European herbalism was influenced by Islamic medicine and in particular Avicenna's *Canon of Medicine* (1025), and later by the influx of new plants from the Americas.

In the Middle Ages monks and nuns, as well as wise women, were the source of plant remedies, and it is often their physic gardens that became the source of invasive plant species, perhaps brought from abroad as precious drugs, their seeds then spreading outside the herb garden, convent or monastery walls.

Hildegard of Bingen, a nun in Germany in the early twelfth century, wrote her own herbal, *Causes and Cures*, based on both the Greek theory of the balance of humours and on her own first-hand observations. She recommends tansy, for instance:

> Reyan [tansy] is hot and a little damp and is good against all superfluous flowing humours and whoever suffers catarrh and has a cough, let him eat tansy. It will bind humours so that they do not overflow, and thus will lessen.[1]

Comfrey is an ancient herb used to heal wounds.

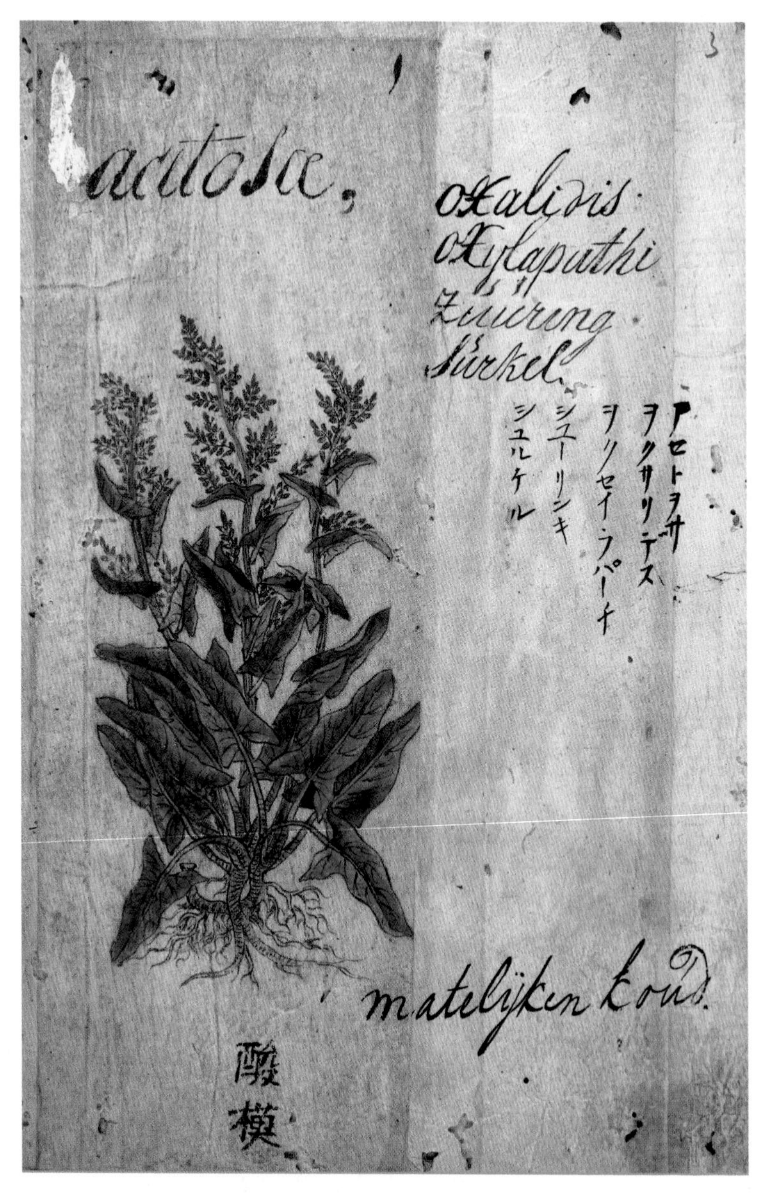

Page from the herbal *Kruid Boek getrokken uyt Dodoneaus* (Herbal extracted from Dodonaeus) with text in old Dutch, Chinese, Japanese and Latin. Rembertus Dodonaeus was a 16th-century physician and botanist.

Herbal remedies were cultivated in monastery gardens.

Tansy is recorded growing in Charlemagne's herb gardens in the eighth century, and in the Benedictine monastery of Saint Gall. Christians were encouraged to eat tansy during Lent, as a spiritual reminder of the Israelites' bitter herbs, but it had a practical advantage too, for it was believed to alleviate flatulence brought on by the Lenten diet of fish and pulses. Sprigs left on a windowsill were thought to repel flies; placed in the bed it repelled bedbugs. A common insect repellent was made from fleabane, pennyroyal, wild garlic and tansy.

Tansy leaves worn in the shoes were thought to prevent malaria, even though the plant is attractive to mosquitoes. With this in mind, tansy is used today in organic gardening as companion planting on vegetable plots. In a similar vein, borage aids the flavour of strawberries and also, like mint, protects cabbages by attracting aphids to itself and away from crops. Tansy was used in the Middle Ages as an aid to conception, though in greater doses served as an abortive. Its oil applied externally was thought to cure dermatitis, and taken internally to kill parasites.

On the other hand, some weeds can have a negative influence on flavour, notably with rue, that 'sour herb of grace', as Shakespeare

A William Bradbury (1800–1869) nature print of a great nettle. The actual specimen was used to create the printing surface, creating an extraordinarily lifelike image.

describes it in *Richard II* (Act III, Scene 4). Yet rue is another all-round cure-all, for complaints of the eye and ear, against hysterics, headaches and fever. Nicholas Culpeper (1616–1654) suggested rue for sciatica and joint pain, and more widely the plant is said to banish household fleas, cure croup in poultry and guard against disease in cattle. The bitter-tasting herb was recommended by Hippocrates as an antidote to poisoning. In the medieval period in Europe it was considered a guard against the wiles of witchcraft and even to offer the gift of second sight.[2] But grown near vegetables you may rue the day, for it is claimed that sage can then become poisonous and cabbages become unpalatable and fail to thrive.

The Chelsea Physic Garden and the records of Culpeper from the 1600s suggest nettles as having a multitude of uses, like tansy: ridding children of worms, easing flatulence and curing diseases of the blood. Iron and traces of copper in the plant proved useful against anaemia. Their sting is said to distract one from melancholy, giving one something else to think about, perhaps. Its dried leaves can be used as an antihistamine, to ease hayfever. Conversely, because of the histamine it contains, whipping with the stems has been a common means of alleviating the symptoms of arthritis. The Romans brought their own nettle seed to the British Isles to help them survive an English winter: 'to rub and chafe their limbs; being told, before they came from home, that the climate of Britain was so cold that it was not to be endured without some friction to warm their blood'.[3]

The forager seeking nettles needs to take courage:

> Tender-handed stroke a nettle
> And it stings you for your pains;
> Grasp it like a man of mettle,
> And it soft as silk remains.[4]

Weeds have always been a source of poison, with field weeds such as darnel causing nausea, impaired vision and ringing in the ears. Deadly nightshade and wild rhubarb leaves are easily confused with spinach, and both can be fatal. John Evelyn's *Acetaria: A Discourse of Sallets* of 1699 contains descriptions of the effects of weed poisoning, and the availability of all manner of cure by herbal remedy, stressing the need for balancing out their various properties:

> We have heard of Plants, that (like the Basilisk) kill and infect by looking on them only; and some by the touch.
>
> The Cooler, and moderately refreshing, should be chosen to extinguish Thirst, attemper the blood, repress Vapours etc.

Deadly nightshade in the wild wood.

Trade card for a cure-all medicine, *c.* 1880.

> The Hot, Dry, Aromatic, Cordial and friendly to the Brain, may be qualify'd by the Cold and Moist; The Bitter and Stomachical, with the Sub acid and gentler Herbs; The Mordicant and pungent, and such as repress or discuss Flatulency (revive the Spirits, and aid Concoction;) with such as abate, and take off the keenness, mollify and reconcile the more harsh and churlish: The mild and insipid, animated with piquant and brisk; The Astringent and Binders, with such as are Laxative and Deobstruct: The over sluggish, raw and unactive, and those that are Eupeptic, and promote Concoction: There are Pectorals for the Breast and Bowels.[5]

The first professor of medicine at the University of Copenhagen, Christian Torkelsen Morsing (1485–1560), saw the study of plants to reduce human suffering as part of his religious vocation: 'To know the herbs and other healing plants which God allows to grow on Earth for the use and benefit of man'.[6]

The Flemish physician Jan Baptist van Helmont believed that God had endowed the earth with all that was needed to cure mankind and that, as with Morsing, it was his duty to use what had been provided. Medicinal herbs were known as simples, as opposed to impositions like bloodletting and purging. The herbals of ancient Egypt, China, India and Europe contain many illustrations and descriptions of

'Simples and Compounds': four pharmacists at work.
Cartoon by 'Pillbox' (Edward Hopley), 1838.

plants for potions and tonics, many of which we now classify as weeds. The 1628 third edition of *The Anatomy of Melancholy*, for example, mentions borage and hellebore, like the nettle, as 'Soveraigne plants to purge the veins, Of melancholy'.[7]

The doctrine of signatures stems from the ideas of Dioscorides, and was taken up by the Swiss physician and botanist Paracelsus (1493–1541). It claims 'Nature marks each growth . . . according to its curative benefit.' In 1621 the German mystic Jakob Boehme published *The Signature of All Things*, and the English botanist William Coles explicitly linked this theory of correspondence with a belief in the Creation: 'the mercy of God maketh . . . Herbes for the use of men, and hath . . . given them particular Signatures.'

Thus plants that resembled various parts of the body were said to be able to treat ailments of the corresponding part. Eyebright was said to cure eye disease. The plant is not mentioned as a cure by Dioscorides himself, nor by Pliny, Galen or Islamic physicians, yet in 1329 it was recommended for complaints of the eyes in Mantua,[8] and by the seventeenth century it was claimed to strengthen the mind and bolster the memory. Culpeper includes a recipe for eyebright:

Studies of eyebright and daisy species. Eyebright is recommended for complaints of the eye.

An Excellent Water to Clear the Sight

> Take of Fennel, Eyebright, Roses, White Celandine, Vervain and Rue, of each a handful, the liver of a Goat chopt small, infuse them well in Eyebright Water, then distil them in an alembic, and you shall have a water will clear the sight beyond comparison.

Milton adopted the cure in *Paradise Lost*, with Archangel Michael giving the herb to Adam after he has been cast out of Eden, the eyebright, or *euphrasia*, appearing to force him to face brutal reality after the Fall:

> . . . to nobler sights
> Michael from Adam's Eyes this film removed,
> Then purged with euphrasine and rue
> His visual orbs, for he had much to see.

Today the notion of 'like curing like' is central to homeopathy, its founder Samuel Hahnemann preaching the 'law of similars' in the first homeopathic hospital in Leipzig in 1832.

It was often left to wise women to dispense plant remedies in the form of potions and poultices. From the late medieval period in the West, before physicians or apothecaries were available to any but the wealthy, women might take on the task of tending illnesses and wounds, and acting as midwives. They were dependent on local knowledge and, if they were literate, on the few herbals available, such as William Turner's *A Newe Herball* of 1568, which had been greatly influenced by Galen (130–201 CE). Balancing the four humours – of blood, phlegm, black bile and yellow bile – was believed to be central to all round well-being. To rebalance an ailing body and mind, Galen recommends, among other cures, sweating and purging, which could both be achieved via herbal dosage.

Poster for a Belgian exhibition on folk medicine. A sickly woman asks for medical help, with superimposed herbs the potential prescription.

For a Christian such as Lady Grace Mildmay in the sixteenth century, nature is innocent of original sin, thus plants and their attributes may come to our aid, to cure or at least decrease the suffering 'hanging over man's head since he let in sickness by sin'.[9] She took on the care of the poor on her estate in Northamptonshire, carefully monitoring symptoms and their possible causes, and the effects of different medicinal recipes. She made up various cordials and syrups, balms, oils and tinctures in alcohol. Distilling opium, she produced laudanum, then a rare drug, and gathered henbane as a source of the alkaloid scopolamine, or Devil's breath, preserving it in ale. Scopolamine has intoxicating and anaesthetic properties, and is today rumoured to

With *gris-gris*, or medicine bags, around her neck, lengths of rag tied with moss and twigs around her waist, and a potted fern and dried leaf on the ledge in front, this herbalist displays her trade. Louisiana or South Carolina, 1929–31.

fuel robberies and rapes in South America. Legend has it that henbane was an ingredient of witches' brew, facilitating broomstick flight, or at least giving one the impression of such. Mildmay's concoctions treated all comers, whether they had epilepsy or smallpox, melancholy or memory loss.

In her kitchen laboratory Mildmay experimented with different recipes, bringing to bear all the influences she had access to at the time, including the ideas of Paracelsus who, contrary to Galen, also suggested mineral remedies and the careful use of poisons, for, he claimed: 'All things are poison and nothing is without poison, only the dose makes a thing not a poison.'

In *A Queen's Delight* of 1671 there is a recipe for a linctus for coughs and consumption, made from elecampane roots, *Inula helenium*, named after Helen of Troy.[10] The plant is said to have sprung from the place where the Spartan queen's tears fell. Pliny mentions elecampane as a medicine, and it is sacred to the Celts who knew it as elfwort. John Gerard, the English herbalist, prescribed it for shortness of breath. It is used in the manufacture of absinthe and today's alternative medicine suggests it as an expectorant, for reducing water retention and for bringing on menstruation.

The 'vertue', or curative advantages, of violets in sugar is similarly wide ranging: 'The heat of Choller it doth mitigate extinguisheth thirst, asswageth the belly, and helpeth the throat of hot hurts, sharp droppings and driness, and procureth rest.'[11]

One of the difficulties in interpreting old herbal recipes is to understand precisely what plant is being referred to. For example, in *A Queen's Delight* invisible ink is to be made from 'fine alum', becoming visible after the writing is held under running water, which may or may not be today's wild alum or wild geranium, and a further confusion is caused by the Linnaean conflation of geranium and pelargonium species. Alum root is a common ingredient in facial astringents. Recently *Pelargonium sidoides*, the South African geranium, known as *umckaloabo*, has been used for colds and flu and even acute bronchitis; it is also being tested as a potential new class of anti-HIV-I agent against the AIDS virus.[12] Yet whether or not modern alum root is the same plant mentioned in the seventeenth-century herbal has to remain uncertain. Similarly, is the plant named 'all-heal' recommended by Theophrastus the plant we know as *Prunella vulgaris* or perhaps *Stachys*, both of the mint family, or might it be valerian? Both are scented plants, and he describes all-heal with the all-encompassing enthusiasm of a Wild West snake charm vendor:

> The fruit is used in cases of miscarriage and also for sprains and such-like troubles; also for the ears and to strengthen the voice. The root is used in child-birth, for diseases of women, and for flatulence in beasts of burden. It is also useful in making the iris-perfume because of its fragrance.[13]

Ground elder is one of the most unmanageable weeds in both Australasia and northern America, where it has been designated an invasive and has been banned. It seems to disappear in the winter season, but its roots and rhizomes creep and prosper deep underground. Just when you think it is no longer a problem it comes up again, in all soil types and positions throughout the growing season, in vegetable plots, herbaceous borders and even in lawns, killing the grass and resisting the mower. Yet it was valued by medieval monks for its medicinal properties, one of its common names being herb Gerard, after St Gerard, who is said to have sent it to monks for the treatment of their gout, for like tansy, ground elder was considered a powerful remedy, its Latin name meaning gout in the foot:

This homeopathy chest contains glass vials and larger bottles, one of which is marked 'Urtica urens', a nettle potion for use on burns and skin irritations. Northampton, 19th century.

Comfrey is an ancient wound herb.

> with his roots stamped and laid upon members that are troubled or vexed with gout, swageth the paine, and taketh away the swelling and inflammation thereof, which occasioned the Germans to give it the name of *podgravia*, because of its virtues in curing the gout.[14]

It was said to serve as both diuretic and sedative. Taken internally it reduced inflammation in joints, and externally the leaves and roots boiled together and applied to the skin when still hot could alleviate sciatica. None of this, however, stopped the plant being invasive, in medieval times as in our own:

> Herb Gerard groweth of itself in gardens without setting or sowing and is so fruitful in its increase that when it hath taken roote, it will hardly be gotten out againe, spoiling and getting every yeare more ground, to the annoying of better herbe.[15]

Another name for ground elder is pigweed, as it is both delicacy and useful medicine for pigs, as mentioned in the Anglo-Saxon medical

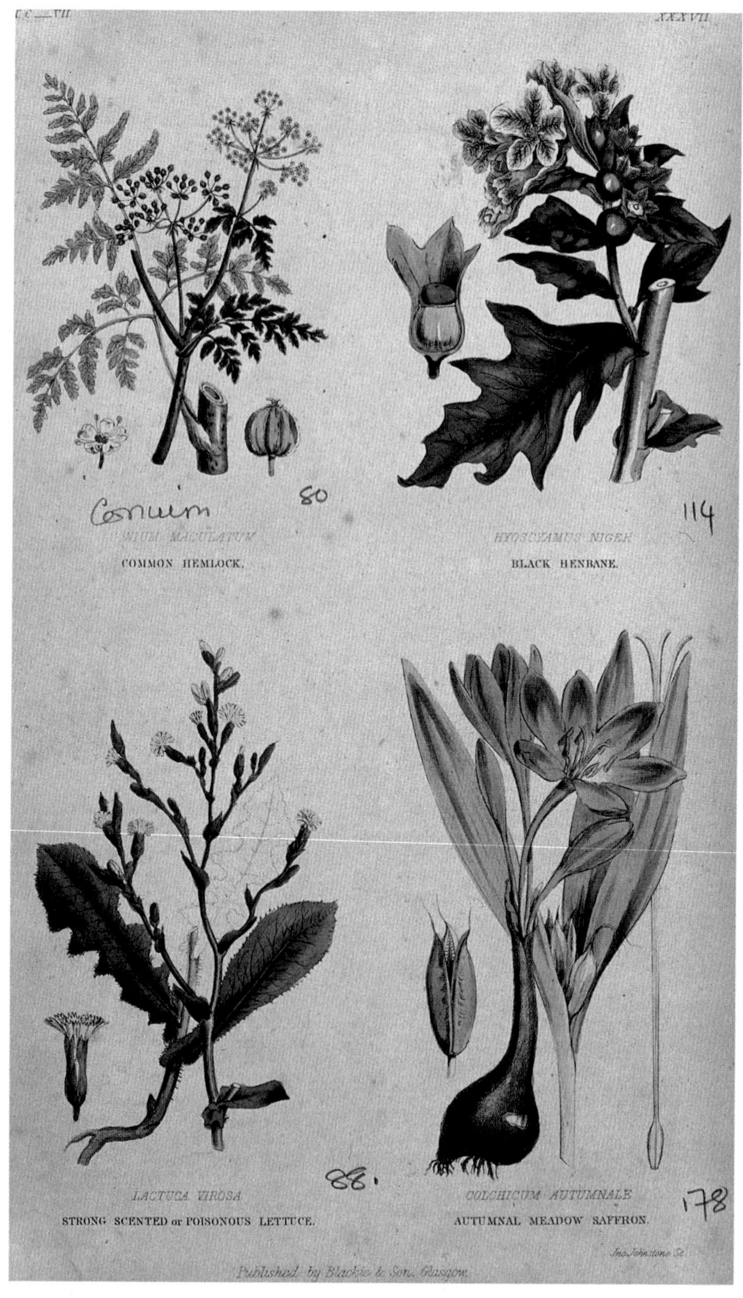

Four poisonous plants: hemlock, henbane, opium lettuce and autumn crocus. Coloured reproduction of a wood engraving by Jonathan Johnstone.

Slicing an opium bud, 1674.

text *Lacnunga*, or *Remedies*: 'To preserve swine of sudden death take the worts lupin, bishop'swort and others, drive the swine to the fold, having the worts upon the four sides and upon the door.'[16]

Early medieval Arabian physicians recognized the medicinal properties of the dandelion, recorded in ancient Egyptian tombs and described by Theophrastus. Its diuretic effects are mirrored in the common names of pissabed and the French *pissenlit*; it is recommended for the liver, kidneys and gallbladder and even for the treatment of diabetes. In India it is also a traditional remedy for snakebites and its milky sap is said to cure surface tumours and warts, and even unsightly moles and freckles.

Yarrow is also mentioned as an effective and practically all-purpose drug, with Achilles in Homer's *Iliad* of 750 BCE applying the leaves to his comrades' wounds. It was recommended by Dioscorides and known as the military herb by the Romans and by the Crusaders as knight's milfoil. It was said to ease haemorrhoids, stop nosebleeds, ease headaches and even unwanted desire, relieving

> swelling of those secret parts, lightly bruised the leaves of common yarrow with Hog's grease, and applied it warm unto the privie parts.[17]

In the field of skincare and cosmetics the uses of plant-based unguents and scents are numerous, from decoctions that are intended to cleanse the complexion and close open pores, to cures for rashes, spots, acne and boils. John Parkinson (1569–1650), the English herbalist, suggested ground elder mixed with cumin seed 'for those who like to look pale'. Chickweed infusion makes a compress for sore eyes and used hot makes a poultice for boils, carbuncles and abscesses. Boiled in lard, chickweed forms an ointment that can soothe insect stings. Parkinson said that nettle leaves if soaked in water make your hair shine and act as an anti-dandruff tonic, as well as protecting against athlete's foot and other fungal complaints. One could rinse one's hair and feet and then use the same liquid as a spray against aphids. A plant like a daisy, unwanted in a lawn, perhaps, is a remedy for sore muscles to the herbalist. The reputation of deadly nightshade – the symptoms caused by atropine from its berries, evoked in the medical student's mnemonic 'Hot as a hare, blind as a bat, red as a beet and mad as a hatter'[18], belies the fact that its name may possibly derive from the Italian women who once used its juice to enhance their beauty by enlarging the glittering pupils of her eyes: a *bella donna*.

The reputation of powerful weeds, able to kill and maim, increases a sense of the otherness of plant life. What may appear harmless can turn out to be deadly as hemlock, a weed used as medicine and poison since 500 BCE and reputed to have ended Socrates' life. Its appearance dissembles, with leaves like an innocent carrot top, and shiny black seeds easily mistaken for aniseed and its roots for wholesome parsnip. Their hollow stems, and those of the water hemlock, have speckled dark blotches known as Socrates' blood, and when fashioned into penny whistles or peashooters, have been known to poison children.

Among her many outlandish attempts to avoid contagion, the woman in this print carries multiple pockets of plants and a basket of potions to prevent infection from cholera, in this 19th-century engraving by Peter Carl Geissler.

In modern medicine plants continue to offer cures. Spurge, or milkweed, which grows wild across Europe, New Zealand and Australia, and is common to North Africa and western Asia, has been found to counter some forms of skin cancer, especially basal cell carcinomas, its very prevalence in cultivated arable land seeming to reinforce its long-used common name, cancer weed.

John Clare noted the many dialect names of wild plants, such as water betony, 'wasp weed . . . a celebrated plant with the gypseys for

'Urticaria': the discomfort of stinging nettles.
Cartoon by 'Pillbox' (Edward Hopley), 1838.

the cure and relief of deafness . . . husk head is the self-heal, a cure for wounds and furze-bind is the tormentil, a cure for fevers, adder bites etc.'[19] Self-heal is a small blue-flowered plant often found growing in lawns, which Gerard refers to as 'not a better wounde herbe in the world'. Culpeper in the seventeenth century declared, 'When you are hurt, you may heal yourself.' Tormentil is used as a cure for diarrhoea in the United States, and the red dye from its roots is still used for dyeing leather. In Bavaria and the Black Forest region of Germany, the crushed roots are a major ingredient in the liqueur *Blutwurz*.

The United States Marine Corps provides training in how to survive when stranded without conventional sources of medicine. Their guidebook recommends certain commonly found algae, for example, as a source of vitamin C and essential minerals. Tea made from burdock root or mint eases a cold or sore throat. Dandelion and rose hip decoctions alleviate constipation; plantain juice, like yarrow, reduces haemorrhoids and staunches haemorrhaging wounds. The unnerving general advice in such manuals is essentially: take a little . . . and see what happens.

John Gerard describes how sixteenth-century field workers in Kent relied on pimpernel flowers as a barometer for the next day's weather. The flowers open only when the sun shines, and close when colder or wet weather threatens. John Clare's 'The Shepherd's Calendar' has the flowers closing 'Which weeders see and talk of rain'. Tiny chickweed flowers also turn their faces to the sun, closing their petals as light falls, remaining closed if the weather is rainy or overcast. In Andrew Marvell's 'The Garden', the flowers are the only true timepiece: 'How could such sweet and wholesome hours / Be reckoned but with herbs and flowers!'

⁂

Weeds are an integral part of biodiversity. Many provide hosts for beneficial pollinators such as bees and butterflies. Charlock, which is

538 DIOSCORIDIS

Millefolium. Digitalis. Primula ueris.

folijs auicularum pennas imitātibus, breui admodū diſſectoque foliorū exortu. folia ſylueſtre cuminum ſimulant, præſertim breuitate atque ſcabritia, breuiora paulò, denſiore umbella, & pleniore. Surculos in cacumine gerit exiguos, & capitula in modū anethi: flores paruos, candidos. Naſcitur in aſperis agris, præcipuè circa ſemitas. Eximij uſus ad ulcera uetera, recētiaque, profluentē ſanguinē, & fiſtulas.

NOMINA ET EXPLICAT.

Græcè Στρατιώτης χιλιόφυλλος.
Latinè Stratiotes chiliophyllos.
officinis Millefolium.
Germanis Garben / Schaffrip / Schaffgarben / & Gerwel.
Galli millefeulle appellant.

Quibuſdam tamen hodie cuminū ſylvaticum, alij anethum ſylueſtre, alij aſininum daucum uocant, ſed falſò ſi uiro doctiſſ. Euritio Cordo credimus qui Stratiotem Chiliophyllon noſtrū eſſe millefoliū in ſuo Botanologico clare demōſtrat, recētiores quoque ad eoſdem uſus, ad quos ueteres. ſuum Chiliophyllum, adhibent, ad ſananda nimirum omnis generis uulnera, diſſoluendum ſanguinem concretum, & ad conſtringendos menſes. Vnde à Stratiote millefolio non eſſe diuerſum omnibus perſpicuum fit.

Verbaſcum. Cap. LXXXIX.

VErbaſcum differētias ſummas habet duas, hoc nigrum, illud album: in quo mas & fœmina intelligitur. Fœminæ igitur folia ſunt braſſicæ, piloſiora multò, & latiora, candida: caulis albus, cubitalis aut amplior, ſubhirſutus: flores albi, aut ex luteo palleſcentes: ſemen nigrum: radix longa, guſtu acerba, craſſitudine digiti. naſcitur in cāpeſtribus. Verbaſcum, qui mas appellatur, oblongius, albis folijs, & anguſtioribus, tenui caule. Sed nigrum prorſus albo ſimile eſſet, niſi nigriore latioreque conſtaret folio. Sylueſtre folia fert ſaluiæ, altas uirgas & lignoſas, & circa eas ramulos quales marrubium: flores luteos, auri æmulos.

Sunt

Yarrow with foxglove and primrose, from Dioscorides, in *De Medicinali Materia Libri Sex*, published in 1543.

Hemlock by a river.

considered a weed in the southeast of the United States, is nonetheless valued by beekeepers because it is resistant to frost, and is therefore a reliable source of early pollen. Nettles nurture many species of butterfly including red admirals, peacocks and tortoiseshells. Conservationists argue that common weeds such as thistles, cow parsley and buttercups can help restore farmland that has been overtreated with weed killers and chemical fertilizers, helping to restore complex interactions between plants, allowing for variations in species and ultimately for an ecosystem that can become self-renewing. On the other hand, weeds are vilified as harbouring pests and diseases inimical to human wants, even though pests that thrive on weeds are often different from those that attack crops. Moreover, invasive weeds are said to threaten biodiversity, particularly in contexts where they meet ideal climate conditions and little resistance; in New South Wales, immigrant species from Europe have decimated indigenous plants and wildlife.

The food writer John Newton suggests that a possible derivation of the word weed may be from woad, since many wild plants were traditionally used for dyeing cloth or camouflaging the skin. Dandelion produces a deep-yellow dye and its roots a dark red; yarrow makes a yellow and drab green hue, used in Russia to colour Easter eggs.

Plant fibre such as from milkweed and yucca can be spun into rope, paper or cloth, and nettle fibre was used in the First World War in Germany and Austria for uniform clothing, when wool and cotton supplies were interrupted by the Allied shipping blockade. In fact nettle cloth had been common in Europe, but from the sixteenth century cotton from India and America had proved easier to harvest and spin, and the Industrial Revolution found cotton fibre easier to machine. The word nettle is derived from the Saxon *noedl*, meaning a needle, which perhaps refers to its sting, but might also suggest its use in clothing. The stems are hollow, which means they can make an insulating cloth, but when the fibre is twisted it can make a finer, cooler fabric for summer heat. Lady Wilkinson, author of a mid-nineteenth-century cultural history of weeds and wild flowers, quotes the Scottish poet Thomas Campbell (1777–1844):

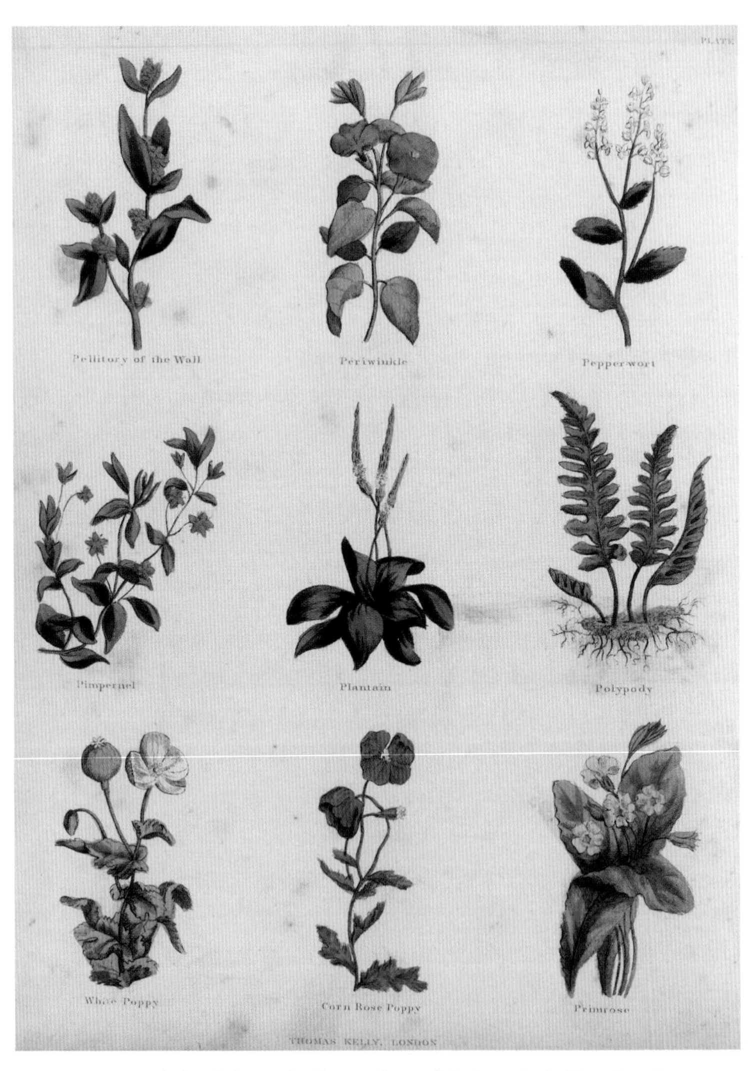

From Nicholas Culpeper's *The Complete Herbal*: plants including plantain, 'with their medicinal and occult qualities; physically applied to the cure of all disorders known to mankind'. Colour engraving, 1850.

> I have slept in nettle sheets and I have dined off a nettle table-cloth. The young and tender nettle is an excellent pot-herb. The stalks of the old nettle are as good as flax for making cloth. I have heard my mother say that she thought nettle cloth more durable than any other species of linen.[20]

She mentions the use of nettle fibre for fishing lines in Kamchatka in the Russian Far East and in northwestern India and Pakistan for a delicate cloth known as *chu-ma*, and that the old German word for muslin is *Nessel-tuch*, nettle cloth. Due to hybrid plants and improved spinning technology, nettle cloth is again in production today.

The legend of the Rhine Castle of Eberstein tells the tale of a maiden forced to spin both her own wedding clothes and a winding sheet for her tormentor, from nettles that grow upon her father's grave, until a kind old woman weaves the plants into garments of extraordinary beauty on her behalf.

Horsetail, according to Gerard, was once commonly used for polishing pewter and thus known as pewterwort. It is abrasive yet relatively gentle, so was also used to scour wood and is effective on modern-day aluminium. A brush can be made by binding together a short bunch of horsetail stems in the middle, so that both ends can then be used. It was also employed in the manufacture of arrowheads. Starch from wild bluebell was once used to glue feathers to arrows. A rush soaked in fat was a poor man's candle. Harvested, dried weeds can be used as insulation and tinder. Dried furze and bracken burn at sufficiently high temperatures to fuel brick kilns.

Chickweed is found in many different terrains and feeds both young birds and the heart and dart moth. Charlock is common in the southeastern states of America and is a virulent weed for crop growers, yet it is prized by beekeepers as a source of pollen. The often cursed burdock is not only a source of a widely used medicine, but its flowers produce a vast amount of nectar, attracting many pollinators. Nettles in fresh and dried form have long been a source of livestock feed. In Scandinavia, Russia and across Europe nettles in hay increase the milk yield in cows and they are enjoyed by poultry and boiled in mash for pigs.

Plant-based compost is a nutrient-rich alternative to chemical fertilizers, even horsetail, for example, which contains silica and can be used to counteract fungal decay and mildew, despite its reputation as a pernicious weed. Compost provides extra nitrogen and improves the soil texture with organic matter.

Planting weeds such as mustard provides living mulch or useful ground cover that can then later be dug into the soil to enrich it, bypassing the composting process and keeping other invasive weeds at bay. Leaving the soil empty over the winter can otherwise cause erosion, in hotter climates creating a dustbowl effect. Green manure is an inexpensive method of soil improvement, particularly if you go to the trouble of saving the seed before digging in, for use the following year. If you are careless and allow the seeds to scatter, then you are creating a further weed problem rather than improving the soil. There are many choices available, depending on how quickly you need the manure crop to grow and the soil type. Blue lupin, for example, has a deep root system that breaks up and aerates heavier soils, allowing for better water retention. It can be used as an intercrop, planted between rows of slow-growing cabbages perhaps, and as a legume, its seeds draw nitrogen in from the air, which in turn feeds other plants. Its blossom invites pollinating bees. Red clover, meanwhile, is a fast-growing perennial whose leaves shelter the soil beneath and is said by some to have medicinal properties, containing phytoestrogens that can alleviate menopause symptoms,[21] and can be used as a salve for eczema and nappy rash.

The types of weeds that grow on a farm or in your garden can tell you what sort of soil you have, and thus indicate what management techniques may be necessary for whatever one wants to grow.

Some liquid weed manures like nettle and comfrey are useful, ironically enough, as a weed killer, and in milder dilutions work against mildew and fungal disease. Like poisonous plants which, used

Flowering stem, rhizome and floral segments of tormentil. Coloured engraving by James Sowerby, 1801, after James Edward Smith.

Phacelia can be used as a green manure.

carefully, are the basis of many effective medicines, weed plants can either take over or if carefully managed can both harbour, protect and even remove other unwanted plants.

Yet for all these examples of the many roles weeds play in nature and in our lives, the foremost use of interest to us has always been as food. We reached out and tried and tested nuts and berries, leaves and roots that grew to hand, testing tender shoots or the juicy fronds of seaweed. We are omnivores and we rely on vegetable food for vitamins and roughage. Weeds have always provided a source of such free food.

six

In Our Diet

Weeds are particularly resilient in dealing with extreme weather fluctuations and major climatic shifts. They often contain a high degree of the essential nutrients on which they and we rely. It follows that eating weeds means eating plants that can offer particularly nutritious food. The naturalist Richard Mabey compares their more 'robust tastes, the curly roots and fiddlesome leaves' to hybrid crops that are 'bland and inoffensive'.[1] Many enthusiasts want to go further, suggesting that eating wild plants forges a connection with the natural world, making not just our physical bodies but our metaphysical being more in tune with nature. Some say it reconnects us with a state of innocence that existed before the modern world corrupted us, when people lived hand to mouth without any notion that they should control what grew.

Few hunter-gatherers remained even in the middle of the twentieth century: scarce Australian Aborigines still following ancient custom in the outback, the pygmies of west-central Africa and a small minority of San speakers in southwest Africa, perhaps. Agricultural and industrial societies have gradually limited their scope, making it difficult for nomadic tribes to move several times a year to new hunting grounds. During the 1960s the anthropologist Richard Lee lived in the Dobe area of northern Botswana with an isolated population of about 1,000 !Kung hunter-gatherers. His findings dispel the idea that the forager's life must have been short and wretched, describing their lifestyle as surprisingly easy-going, with men hunting prey and women

still wet-
unripe berries
Aug 31

Autumn leaves and fruits of bramble.

gathering plants. Neither sex worked long hours and they tended to live to ripe old age. Moreover, it is a mistake to think that foragers never made any attempt to control their food supply. Both in East Africa and the west of North America there is evidence that grassland was periodically burned in order to generate new growth, and thus attract game.

Richard Lee found the foragers he was studying ate mongongo kernels as an important source of vitamins, the trees growing on desert sand dunes.[2] Mongongo trees carry small reservoirs of water in the crotches of their branches, which meant that the plant could also quench the foragers' thirst. Like many land foragers, they survived mostly on vegetation, fruit, nuts and insects, as game was often hard to come by. The charity Elephants Without Borders noted that elephants love to eat the mongongo fruit, but their digestive system is unable to break down the extremely hard shell and obtain the meat within. The !Kung follow the elephants to collect the nuts, the digestive process having softened them sufficiently to make them easier to crack open – and the shells and dung are a useful source of fuel.[3]

Those hunter-gatherers surviving on an aquatic larder – of fish, marine mammals and seaweeds – as a more constant and reliable source of nutrition are distinct from land foragers in that they are able to settle more permanently, as on the Aleutian Islands of Alaska.

Foremost in advice given to foragers for wild and weedy foods today are instructions to avoid confusing the poisonous with the innocuous plant, and to look out for the danger signals of the overripe or the almond-scented. Poison ivy can all too easily be mistaken for mango or sumac; hemlock for tasty wild carrot or parsnip. Even if plants are correctly identified, gatherers are well advised to avoid urban weeds that may have become contaminated: trace levels of lead, arsenic and mercury may lie hidden in the soil, so that a salad of young nettle leaves plucked from a graveyard should be avoided, while weeds flourishing on roadside verges may have become polluted by emissions from passing traffic. Those collected from the edges of fields may be tainted with pesticides; passing dogs may have fouled low-growing woodland

Jacques Nimki's *Black Begs*, 2010, digital drawing.

plants; cats are attracted to the smell of the dandelion, said to remind them of tomcats, and have a tendency to spray its leaves.

The foraging of the past, which has always existed in less developed countries, has lately become almost de rigueur in the West. Beliefs surrounding the protection of our ecology from the damage for which we should take the blame assume that nature would be fine without our interference. Ecologists are concerned when governments begin to take biofuels seriously only when oil reserves are running low. This polarization of views has unfortunate consequences; for example, because seaweed might be a viable future source of ethanol, those who might otherwise have supported its consumption as a healthy food are suspicious that seaweed gathering might lead to ethanol production, and are concerned to protect coastlines from future fuel-plant pollution.

While coastal communities have always eaten seaweed, in recent times its consumption has dwindled in the West, apart from a few iconic dishes such as Welsh laver bread, made from laver and oats, or sweet Irish seaweed pudding, made with carrageen moss, which is

A leafy green tunnel of weeds.

also a source of a thickening agent in many products such as salad cream, ice cream, toothpaste and even paint. However, seaweed still plays an important role in much Asian food. In China enormous quantities of seaweed, or *gim*, are gathered, even though levels of toxicity in China's waters can be high, carrying through into the weed. If you want to gather seaweed, it is advisable to steer clear of waters near nuclear power stations or industrial plants.

Dried kelp is widely used as a salt substitute and has the advantage of containing added vitamins and minerals. Agar-agar, a jelly derived from sea algae, not only served as a wound dressing in the First World War, but when added to food is a source of iodine which can help with thyroid problems. In Belize a traditional drink is made from seaweed and milk, spiced and sweetened, called *dulce.* Korean and Japanese

A dandelion by William Kilburn, 1777. Kilburn produced many of the plates for William Curtis's *Flora Londinensis*.

traditional dishes incorporate all manner of sea vegetables, and despite seaweed's tendency to soak up available toxins, it is considered, along with fish and seafood, a very healthy diet. Sheets of dried seaweed or *nori* encase rice balls and sushi, fried strips ornament seafood broth and hotpots, and fleshy, subtly flavoured *wakame* has been a prized ingredient of salads since 700 CE – and yet it is listed as among the

Junks laden with seaweed in Fusan harbour, Korea, 1904.

world's 100 most dangerous weeds in the Global Invasive Species Database of 2009.

To return to land-grown weeds, the environmental activist Wendell Berry's assertion that 'eating is an agricultural act' marks his desire to integrate farming with the natural world.[4] He describes the 'little patch of weeds' that a person might hanker for as representing home, 'holding it bright in memory, and love the saplings and the weeds'. Arguing that weeds are part of an interconnected food system, he allowed giant ragweed and jimsonweed on his Kentucky farm, controlled by carefully timed mowing and grazing with sheep. In like manner the wildflower meadow at Great Dixter in East Sussex, owned by the late gardener Christopher Lloyd, is cut late in the season, so as to allow a wide range of seeds to ripen. The hay is left on the ground until the seed is distributed, before the grass is cut again short to over-winter; other areas are left uncut, in order to provide winter cover for other creatures.

❧

A fear of genetically engineered foods has increased interest in organic food. Contemporary foragers claim that wild food is two or three times more nutrient-dense than anything grown as crops. Phrases like the 'spice chest of the hedgerow' suggest an abundant resource that is available to all, but the amateur gatherer has to learn how to distinguish edible and non-edible plants. Wild garlic sprouts, for example, have the appearance of poisonous lords and ladies. The food forager Robin Harford would have us treat this random wild larder with a certain delicacy, cautioning against clumsy greed: 'gather tenderly, as you would stroke a lover. Stand back, start slowly, do not rush in grabbing as much as you can in your fear of scarcity.'[5]

❧

In the seventeenth century John Evelyn's *Acetaria* compared different countries' tastes, with Spaniards and Italians said to enjoy ox-eye daisies in the spring and wild garlic at all times: 'To be sure, 'tis not for the Ladies Palats, nor those who court them.' French country folk are reported to adore dandelion roots; the Genoese white poppies. Horseradish is mentioned, and he regrets classical salads that have fallen out of favour, such as Emperor Nero's favourite silphium, prized for its medicinal properties: 'A wonderful Corroborater of the Stomach, a Ristorer of lost Appetite and Masculine Vigour . . .'.

Advice on how to distinguish safe foods and minimize risk often suggests testing the different parts of a plant separately. Some weeds should always be treated with the greatest caution, such as hogweed, which can lead to painful blisters even if cooked. Generally speaking, if in doubt, then strong scent or an initial bitter taste are important warning signs. One might just say, if it tastes foul, then spit it out. The

Marine Corps' Universal Edibility Test advises not eating for eight hours before testing an unknown plant, which might be awkward in embattled conditions. One should first of all check for reactions by rubbing a leaf on the skin of the inner elbow. After this one might rub a leaf against one's lip, wait fifteen minutes, and then put a small piece within the mouth. If there is any sensation of burning, numbing or stinging, one should immediately abort. If all seems well one can eat a small portion, wait another eight hours, and if any reaction is noted then immediately induce vomiting. The test suggests that at first one should check that the plant in question is sufficiently prolific to make such extensive and time-consuming testing worth one's while.

Plants that can be edible once cooked are sometimes dangerous raw, so that, for instance, ground ivy is tasty if stewed but should be avoided in a salad. It was once used by beggars in the Middle Ages to create sores, by rubbing its sap into the skin. It contains protoanemonin, an acrid sap that is destroyed in cooking.[6] Oxalic acid occurs in wood sorrel, which can be a contributory cause of kidney stones, and riverside Himalayan balsam is also packed with oxalates that similarly are destroyed by cooking. Cuckoo pint bears poisonous berries and the plant has a noisome odour, but if you do wish to eat its leaves, they need to be well cooked to remove all toxicity, otherwise its effect is of needles being repeatedly forced into the tongue.[7] Burdock leaves are bitter tasting, though the Japanese have been known to eat them when the plants are young, as a spinach substitute. The roots are more palatable, but they contain a fibre known as inulin that is indigestible, and when it ferments 'inside the insides' it can cause flatulence. Chickweed is tasty, but its stringy stems can easily get caught between the teeth.

Foraging for wild mushrooms is almost by definition a foolhardy activity unless you are well informed. The fairy-tale and folklore potency of enticing spotted red toadstools that poison at a single bite are matched by real-life evidence of the extreme toxicity of fly agaric, for example. Its careless consumption can be a fast-track to a liver transplant. Names like death cap and destroying angel are good

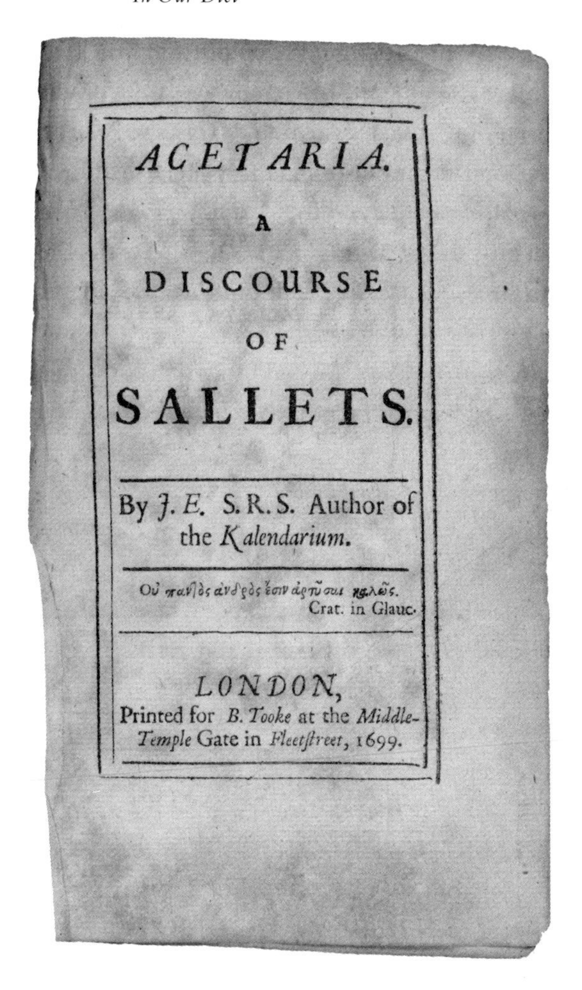

ACETARIA.

A

DISCOURSE

OF

SALLETS.

By *J. E.* S. R. S. Author of
the *Kalendarium.*

Οὐ παντὸς ἀνδρὸς ἐστιν ἀρτῦσαι καλῶς.
Crat. in Glauc.

LONDON,
Printed for *B. Tooke* at the *Middle-Temple* Gate in *Fleetstreet*, 1699.

The title page of John Evelyn's *Acetaria* (1699).

warnings, but you still need to be able to recognize them from edible fungi. Morels and puffballs harbour microscopic insects and grubs, so need to be soaked overnight in salty water before consumption.

Despite these pitfalls, dangerous and less so, eating wild plants is growing in popularity in the West, at least among those of means. Look in any cutting-edge produce market and you may find the very weeds you have been assiduously removing from your garden. Purslane is in vogue, even though it is considered a weed in many countries, famed for its antioxidant properties and high levels of omega-3 fats, and thus considered a healthy choice, inhibiting cancer risk and heart disease. In a 100-g (3½-oz) portion 'you can get more than 1,320 international units of vitamin A, 21 mg of vitamin C, and a dense array of B-complex vitamins'.[8] It has also been praised for its delicious, delicate lemony taste and the crunchy texture of its leaves. The plant is native to India and Iran, but now grows all over the world; it is used in little baked pastries in Turkey, fried with other vegetables and feta in Greece as *andrakla* and forms the major ingredient of a Portuguese soup, *baldroegas*. Pliny the Elder suggested it should be worn as an amulet to ward off evil and Culpeper suggested that the leaves somehow fasten loose teeth. In *Acetaria: A Discourse of Sallets* (1699) John Evelyn describes purslane as 'generally entertained in all our sallets. Some eat of it cold, after it has been boiled, which Dr Muffet would have in wine as a refreshment.' The Californian chef Christopher Kostow dresses young purslane simply with a pickled lime vinaigrette.[9] He argues that older plants need to be 'denatured', dressing them with *vincotto*, a syrup made from grape must, to tenderize the stalks, and

Twenty species of fungi, including fly algaric and the death cap. Coloured lithograph by A. Cornillon, 1827.

since they can 'stand up to strong flavors', sometimes serves them with nectarines and plums, toasted hazelnuts and even blue cheese.

Carlo Mirarchi, a Brooklyn chef, in a similar vein dresses strongly scented and flavoured lovage, an important medicinal plant in the fourteenth century, with nut milks, to take away its acidity. He argues that such an unusual combination invigorates the palate. When you have a salad like that, it is 'like you're some woodland creature jumping around eating all this weird stuff'.[10] The British chef and real food campaigner Hugh Fearnley-Whittingstall recommends lovage instead of parsley or celery, though he warns that it is punchier; it is termed *celery bâtard*, or bastard celery, by the French. He uses it to stuff fish and chicken and to flavour a summer soup of lettuce, pea and cucumber.[11]

Michael Pollan pairs lamb's quarter with purslane – as two of the most nutritious plants in the world. Lamb's quarter is an annual, eaten by many wild mammals, birds and insect species and producing thousands of seeds, which can be harvested and ground into a dark flour for bread. Its dried leaves also make a flour, better suited to flatbreads. It is good for salads, stir-frys and stews, with a rich, mineral taste not unlike that of chard.

A pleasant way of trying to keep ground elder under control is to pick its youngest shoots throughout the growing season, before the leaves have unfurled, and fry them simply in olive oil, but then many things are tasty fried and served with salt. However, the hardy perennial is likely to benefit from such top pruning, its roots becoming stronger, taking encouragement to send up ever more shoots, its network below the soil surface severely inhibiting other plants. And yet a summer quiche full of its tender petioles is a fine thing:

Ground Elder Quiche

100 g (3½ oz) shallots (sliced)
100 g (3½ oz) blanched ground elder
3 eggs
150 ml (5 fl. oz) plain yoghurt

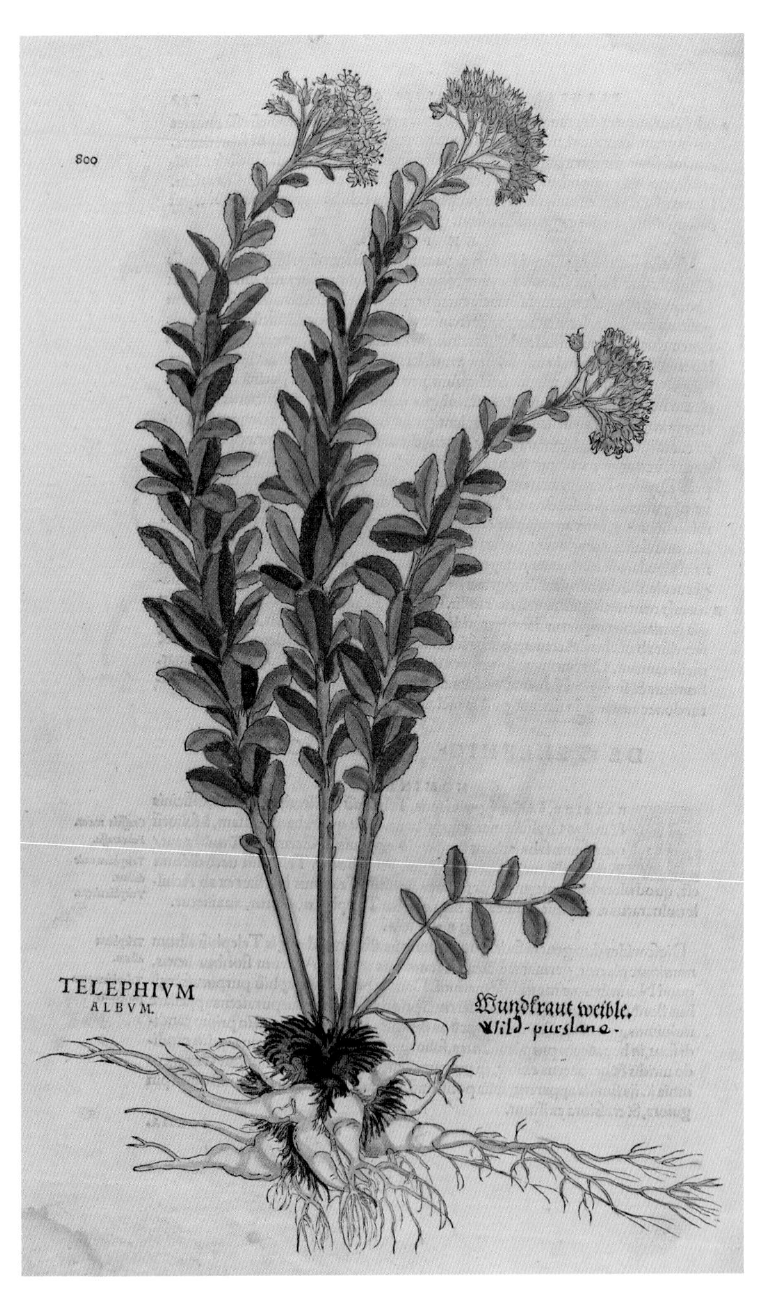

Wild purslane, from Leonard Fuchs's *New Herbal* (1543).

150 ml (5 fl. oz) whipping cream
50 g (1¾ oz) grated cheese
nutmeg
salt and pepper

Fry the onion until just translucent, then add in the ground elder leaves, stir and fry for a further minute. Whisk together the eggs, cheese, yoghurt, cream, nutmeg, salt and pepper. Line a buttered 26 cm quiche tin with pastry and pour in the filling. Bake for 30–35 minutes in a medium hot oven until it is risen and golden.[12]

Willow grows wild in Alaska and is a traditional food, eaten as young shoots in its raw state, and cooked with meat or soaked in seal oil in the Eskimo manner, when it is known as *pahmeyuktuk*. Great willow-herb flower petals are used to flavour a sweet jelly, eaten with cold meats or simply spread on bread.[13]

Few weeds are simpler to identify or more available than the dandelion – its bold, golden flowers above all – so it is easy to gather them to make fritters, and picking them in the garden before they form clocks stops them from spreading:

Dandelion Fritters

1 egg
125 g (4½ oz) flour
250 ml (9 fl. oz) milk
dandelion flowers, snapped off with a short stem remaining
vegetable oil

Holding the flowers by the stem, dip in a batter made from the egg, flour and milk and fry them in oil, draining on a paper towel. Season with salt and pepper to taste, or roll in icing sugar.[14]

John Evelyn praised the culinary value of nettles, 'the Buds, and very tender Cimae, a little bruised, are by some eaten raw, by others boil'd, especially in Spring-Pottage, with other herbs'.[15] In 1661 the diarist Samuel Pepys remarked on a meal provided by his friend, Mr Simons: 'There did we eat some nettle porridge, which was made on purpose to-day for some of their coming, and was very good.'[16] Nettles are an excellent source of iron, calcium and magnesium and many vitamins. Lady Wilkinson informs us that they produce a rennet for coagulating milk, and one which has less of an aftertaste than ordinary rennet, which is taken from the stomach lining of calves. Nettle rennet is thus suitable for vegetarians. Make a simple pesto from nettle and other weed leaves, such as hairy bittercress and wild garlic, enrich with wild nut kernels, and the result can be surprising, transforming a dish of pasta.

Meadowsweet makes a rich milk jelly, served with a cherry sauce, and rosebay willowherb petals are added to a scone mix by Gail Harland;[17] crab apples are sweetened with honey and the angelica flavour of wild lovage. Foraging lends itself more easily to sweet desserts, perhaps, because we find it easier to identify what is safe and wholesome.

Wine can be brewed from many weeds including nettle, dandelion and elder, the latter from the berries and sparkling wine from the flowers. You can prick sloe berries, cover them with sugar and steep in gin for a rich, jewel-like liqueur. Various seeds and toasted roots make ersatz coffee and tisanes are brewed from many hedgerow herbs.

I recall dandelion and burdock 'pop' from childhood, straight from the adventures of Enid Blyton's *Famous Five*, conjuring up midnight feasts and dangerous escapades, being somehow manly, imbued with notions of camping and the great outdoors. Originally, in the Middle Ages in Britain it was used for a mead, brewed from the roots of both plants with honey and wine. Gradually it became used for a non-alcoholic drink, with a taste not unlike sasparilla. Though shop-bought versions today contain little or no burdock nor dandelion root, even in its denuded, commercial form the name alone seems to lend it an aura of healthy ruggedness. Here is a recipe for the real thing:

Dandelion and Burdock

600 ml (2½ cups) cold water
1 tsp ground dried burdock root
1 tsp ground dried dandelion root
small piece ginger, crushed
1 whole star anise, crushed
½ tsp citric acid
300 g (10½ oz granulated sugar)
soda water

Place all of the ingredients, except for the sugar and soda water, into a large saucepan and bring to the boil, and simmer for 20 minutes. Filter the mixture through a sieve lined with a tea towel or muslin cloth. While still hot stir in sugar to taste, until it dissolves. Leave to cool. To serve, add 200 ml (7 fl. oz) of soda water to every 50 ml (1¾ fl. oz) of syrup and stir well, then pour over ice in glass tumblers.[18]

In Sweden yarrow was known as a 'field hop' and used in brewing. If you prefer a drink with more punch, then henbane, or killer of hens, when added to beer in the Middle Ages greatly increased its inebriating effect. So prevalent was the custom that in Bavaria in 1516 a law was passed to allow only hops, barley and water in beer, though yeast was later permitted. Given henbane's reputation as an anaesthetic, one can imagine the effect it might have had, particularly when large quantities were added.

There is nothing new in the idea of wild, disregarded plants being edible and highly nutritious, and even full of flavour. In the developed world today, to hunt for weeds for the table has become a recreational activity. Apart from a few scattered minorities, we are no longer hunter-gatherers scouring the remaining wild countryside for sustenance. Yet the pleasure we may feel in finding a particular plant can be as enjoyable as a treasure hunt. As allotment holders or gardeners, the vegetables we grow may seem relatively under our control, and what we find we cannot control is nominated as a pestilential weed. Michael Pollan talks of hunting for our food, rooting out scarce fungi in the woods and trying to avoid all their cunning, deadly cousins that lie in wait to trick us, as teaching us 'something about who we are beneath the crust of our civilized, practical, grown-up lives'.[19] This sort of enjoyment can imbue the eating of wild weeds. The language of the forager calls garlic wild to make it seem more exciting, and there is an emphasis on freshness. Foraging is often described as an adventure, as revolutionary and inspiring.

A *veldfood* (wild food) project in South Africa sets up organic vegetable gardens that attempt to include only indigenous planting: a garden of cultivated weeds, if you like. The long-forgotten foraging plants of the Khoisan are prepared in Shaun Shoeman's restaurant

Elderflowers are used to make cordials and sparkling wine.

A salad of nettle, dandelion and wild rocket leaves with green alkanet and buttercup flowers.

near Franschhoek in Western Cape Province.[20] Shoeman's recent ancestors would have gathered these same plants as wandering hunter-gatherers. Apart from their role in the kitchen, he has high hopes that the lemon-flavoured *buchu* leaves, the wild garlic, *tsamma* and *makataan* melons and *num-num* plums can 'unite cultures and help South Africans not only to understand each other better, but also to take pride in their heritage'.[21] This is the cry of many who believe in the value of eating weeds.

seven

A Wild and Weedy Garden

On a visit to the Imperial Gardens in Tokyo I was struck by the sight of a group of elderly Japanese tourists taking close-up photographs of the weeds that had been allowed to remain in the manicured gravel. They were peering at a single spear of barnyard grass. It was 3 centimetres high or so, smooth and green with a reddish base. Its emerging leaf was still coiled, narrow and smooth on both surfaces. I noticed these details because they were noticing. I recognized what it was only because in the city where I was living at the time, where small urban paddy fields were tended by my neighbours as tenderly as the most prized rose garden in the Home Counties, it had been pointed out to me as a hated weed and I had looked up the English name and saw that it grew all over the world. This is the weed mentioned in chapter Four, that is the scourge of rice paddy fields, but is quite happy to grow with its feet out of water too.

In my provincial Japanese city the elderly ladies would spend long hours hand-weeding their plots, in sun and rain, in their woven conical straw hats or *sugegasa*, their backs bent in the humid summer, to the tune of a million frogs. These fields did not seem to be grown out of economic necessity. Small groups of women joined together to grow fresh food, and it seemed a culturally important pastime, reminding them of their youth, perhaps, and the relation between themselves and the land.

Yet that day, on what may well have been a once-in-a-lifetime visit to their capital, a photograph of this noxious weed seemed to strike them as a worthy memento. Here was *multum in parvo*, much in a little weed. In this new context it seemed no longer to be scorned, or perhaps they delighted in its prevalence and quite simply it reminded them of home. All the same, I could not help wondering if the plant would be permitted to spread its seed. Would the imperial gardeners be out next spring, trying to eradicate a surprising spread of barnyard grass with their tiny hand-held hoes?

The idea of a garden allowed to grow as it will, without human interference, has become fashionable again. As soon as technology allowed us the possibility of a weed-free garden then the notion grew up that this might be a pity. Something might be lost *sans* weeds. The implicit question is whether a garden might be more aesthetically pleasing if the struggle were set aside. James Shirley Hibberd, an influential nineteenth-century British gardening writer, epitomizes the view that the gardener must marshal their forces against any such non-interference: 'Gardening is always more or less a warfare against nature.' On the one hand one might argue that a garden is, by definition, managed ground, but on the other there remains a common longing for a piece of untrammelled nature, where we allow it as much freedom as we dare, without some inconvenient plants, some particularly audacious weeds. Weeds without what Michael Pollan aptly calls their ecologically sophisticated quotation marks.[1] Would such a garden be a natural garden then, or a 'natural' one? Perhaps it has to be even more managed than the more obviously designed gardens, with straight lines, neat parterres and clipped hedges?

The idea of a wild garden, a designed, manageable form of wilderness, was first suggested by the Irish gardener and writer William Robinson, in a seminal book of 1870, *The Wild Garden*. In the magazine *The Garden* in 1872, he laid out his theory:

> A lady correspondent asks us what we mean by the term 'wild garden,' which was new to her. The Wild Garden is one where we plant, but do not mow, or rake, or trim, or stake; and wild gardening simply means the substitution of beautiful

The more formal part of Gertrude Jekyll's restored gardens at Upton Grey, overlooking the planned 'wild'.

A woman weeds a vegetable bed in Camille Pissarro's *The Artist's Garden at Eragny*, 1898.

> hardy plants for the weeds and brambles which cover such a comparatively large surface of the ground near every country seat . . . there are at least five hundred kinds of ornamental hardy exotic plants that will thrive perfectly, instead of the weeds . . . in all the rougher parts of our pleasure gardens.[2]

Trained in the formal gardens of the Curraghmore estate, and later at Ballykilcavan, he advocated doing away with the bedding out that was so much a feature of nineteenth-century planting. It seemed to him to be too time-consuming and to create a 'repulsively gaudy' effect. It was not just native wild plants he used, but whatever he encountered abroad on his travels that seemed to offer both beauty and the ability to grow unaided in different conditions. On wind-blown, arid soils he recommended alpine flowers, for example, after observing them in their native setting in the Swiss and Italian Alps. This openness to alien species, and possibly a lack of foresight, caused him to champion both Japanese knotweed and giant hogweed. Robinson wanted gardens to appear less designed, or perhaps seem closer to God's original design. Plants, whether native or exotic, seemed more integrated if they were surrounded by what had previously been rigorously pulled up. If the planting was sufficiently well planned then it would manage to keep any incipient weed expansion at bay. Only weaker weeds, or spray, were acceptable:

> To most people a pretty plant in the wild state is more attractive than any garden denizen. It is free and taking care of itself, it has had to contend with and has overcome weeds which, left to their own sweet will in a garden, would soon leave very small trace of the plants therein; and moreover, it is usually surrounded by some degree of graceful wild spray.[3]

The interest aroused by imported plant species, and the sense of competition to grow new exotics, is often used to explain the success of many invasive weed species: 'Botanists would rather receive one of our most common weeds from a newly-discovered, or newly-explored country, than a new species of an already known genus.'[4]

Gertrude Jekyll, an English garden designer renowned for her painterly mixed-flower borders and her belief that no soil should be visible in a summer border, was prepared to use all comers that might suit her palette. She sought to create the illusion that plants had just happened to occur in the subtle colour combinations she designed: 'There is no spot of ground, however arid, bare or ugly, that cannot be tamed into such a state as may give an impression of beauty and delight.'[5] In the gardens of the Manor House at Upton Grey in Hampshire, first planted by Jekyll in 1908 and now restored to a seemingly effortless glory, there are areas of woodland that appear convincingly natural, with drifts of spring bulbs running through grasses and wild flowers. Paths are mowed through the meadow planting, accentuating their profusion. But such apparent randomness requires close attention or it will quickly be taken over by the more hardy, spreading weeds. It is a conceit: a paradise vision of plants in mutual abandon.

More recent wild gardens aim to invite insects, bees, hoverflies, small mammals and amphibians in, as if all but man's presence will enhance a sense of natural order. *Vogue* magazine calls a stylish contemporary garden 'not so much overgrown as it is untamed'. The gardener in question claims that it is her intention to make the plants seem like weeds: 'I teach the plants to become wild again.' And this version of wildness is intended to suggest the playful: 'I encourage things to become a weed, because they're all weeds somewhere in the world.'[6]

There is interest now in seeing weeds as a means of protecting and re-establishing endangered wild plants and creatures. Plants now designated as weeds have adapted to survive in man-made environments. The same plants may grow in wild places, away from our influence, where there is no need to describe them as weeds, but they must retain their weedy qualities if they are to be plentiful, as opposed to rare and delicate plants. Many continue to survive only because we have provided the environments in which they can thrive, through farming and gardening, and through the microclimates of our cities.

Jean-Baptiste Oudry's study of the overgrown, abandoned gardens of the Prince de Guise's château at Arceuil, France, *c.* 1744–7. In the foreground the weeds grow without restraint, conveying a sense of melancholy.

Caruthers on a grass path cut through the wild.

What evidence is there for the continued significance of weeds in our lives? There is the real food movement and a resurgence of interest in herbal remedies, encouraged by increasing human resistance to antibiotics. The idea of wild flower meadows offers us comfort, even if we never have anything to do with them. Increasingly there is a desire to allow weeds to flourish both in the domestic garden and in areas such as motorway sidings, beside rivers and canals, and farmers may allow a percentage of their fields at least to be left untreated by herbicides. Organic farming methods entirely prohibit chemical weedkillers, so that troublesome invasive plants have to be carefully managed, using traditional, natural methods of control.

Archaeological farm projects around the world reveal early farming methods. Lyon Farm in DeKalb County, Georgia, provides evidence of creek settlement and farming prior to 1800 and there are projects under way to understand more fully the ancient methods

overleaf: *A Formal Garden* by Johannes Janson, 1766. The *broderie parterre* (embroidered flower bed) is filled not with flowers but with contrasting soil and gravel, neatly framed with small box trees, as if even cultivated plants might not express sufficient sumptuous distance from what grows naturally.

In Claude Monet's *The Artist's Garden at Vétheuil*, 1880, the grass has been allowed to grow long and the beds are crammed with common sunflowers.

of Neolithic farmers in Denmark. The Butser Ancient Farm Project in Hampshire has reintroduced cornfield weeds in one of its crop sites and grows Bronze Age crops that have since been considered weeds, like wild carrot and fat hen. Studies of West African weeding practices among the Dogon of Mali may hold the key to low-cost, low-risk weed control.

The body of knowledge about farming without herbicides questions the assumption that weed-free fields are best for global ecology. Set-aside policies in the European Economic Community, for example, though set up to reduce surpluses, have had the knock-on benefit of mitigating damage caused by intensive farming methods. Biofuels are being developed from quick-growing weed species, in an attempt to provide new sources of energy in light of soaring oil prices, and arguably offering some response to global warming.

The urban green roof movement celebrates roof gardens, urban balconies and even planted walls with built-in habitats for wildlife, laid out in such a way as to seem as if plants had happened to grow there accidentally, in the interests of biodiversity but also of the psychological benefits they entail. Conservationists have advocated guerrilla tactics, creating pop-up gardens, illicit temporary green sites on disused ground or filling spaces left after building demolition and before a new one is built. Such gardens are often slap bang in the city, growing tomatoes and green beans, marigolds and giant sunflowers, in an ethos of anything goes. Guerrilla gardens have been turning up all over the developed world, in Auckland in New Zealand, Bologna in Italy, Lublin in Poland and Graz in Austria, to name just a few.

One garden has grown up on a traffic island on one of the busiest road junctions in London, at Elephant and Castle, an area where dilapidated, brutalist architecture dominates. The High Line, a rail line on the West Side of New York, has gradually been turned into a city-backed park, 'inspired by the self-seeded landscape that grew on the out-of-use tracks during the 25 years after trains stopped running'.[7] The plants

In the major road junction at Elephant and Castle, London, guerrilla gardeners confront the traffic with sunflowers.

Tumbleweeds line the open highway in Shannon County, South Dakota, 1940.

that were chosen to create this winding green ribbon of vegetation were chosen for their hardiness and sustainability, or, one might say, their weediness: vetches, milkweed, cranesbill, spurges, sumacs, hyssop and loosestrifes, all stalwarts of the weed family.

Many conservationists suggest that just leaving roadside verges and central reservations to grow, or taking care to cut them back at the optimum time after seed production, will create safe habitats for many categories of wild species. Britain's wayside verges alone amount to thousands of acres of potential wild meadow. Public authorities respond that it is not merely a question of untidiness but more importantly lush vegetation may obscure the sightlines of passing vehicles, leading to more accidents, though this problem is solved by cutting a swathe of turf closest to the road. Conditions in these corridors of vegetation can create their own microclimate, and seeds are easily dispersed and spread by the turbulence of passing traffic. Highways pass through both undeveloped terrain and all manner of human construct, farmland, industrial plant, suburb and inner city. According to a Cambridge University blog it is reported that scurvy grass, which

An apothecary bottle for scurvy-grass water, used as a cure for scurvy and scrofula.

An allotment overtaken by weeds, fast returning to wild meadow.

usually grows in very salty soil, on salt marshes and beaches, has been found growing vigorously on inland roadside verges.[8] The increased use of gritting during the winter months has resulted in a build-up of residue salt ideal for the plant to prosper. Scurvy grass is edible, related to the cabbage family, and contains high levels of vitamin C, essential to the good health of sailors on long voyages, hence its common name. Unfortunately it also has a pungent flavour, and its location might anyway be off-putting for the table, coated with carbon monoxide and diesel fumes.

The Urban Heat Island, or UHI, effect of our cities, is exacerbated by a lack of vegetation cover, with so many hard, reflective surfaces multiplying the heat of the sun, such as the recent tendency in the suburbs to asphalt over what were once front gardens to create free parking. Whitened roofs, pavements and all hard surfaces can increase the albedo of built-up areas, and should be coupled with the creation of green space wherever possible, exemplified by many guerrilla gardening projects. Even in countries that we think of as relatively cold, such as Canada, these strategies are advised when, at the extreme, overheating has been reported as causing fatalities.

The question is whether or not we face extinction, from global warming and from the spread of disease amplified by global travel. The continued decrease in animal and plant species might herald a world with only couch grass, dandelions, ground elder, with rats, cockroaches, pigeons . . . and with humans just about clinging on. In this apocalyptic vision, it has been argued that 'the species that survive will be like weeds, reproducing quickly and surviving almost anywhere'.[9] Those who do survive will need to be 'scrappers, generalists [and] opportunists'. David Quammen posits that we may escape complete extinction. He argues for a new Eden, a luxuriant garden of new and unfamiliar plants, when *Homo sapiens* will also be transformed, when 'wondrous forests are again filled with wondrous beasts. That's good news!'[10] This latter epithet is surely ironic, a prediction of this new future unknown form of humanity, comparable to the promise of a replacement loved one, after someone's death.

To return to weed basics, to weed requires close attention to detail, or you risk pulling up a desired plant. This single-minded concentration can create another sort of hotbed environment where the initial groundwork for human species fertilization can take place. Some American farms, no doubt looking for fresh opportunities to increase their income, run weed-dating sessions. In Idaho, Earthly Delights farm helps singles meet up over a spot of intensive weeding and between the rows of lettuces, strawberries and tomatoes love can sometimes bloom.

It also gives weeding a helping hand.

Chandelier in the *Fragile Future* series, by Lonneke Gordijn and Ralph Nauta of Studio Drift, 2011. Dandelion seeds, combined with lighting technology materials of phosphorous bronze, electronics, LEDs and Plexiglas.

Afterword

A common response to the idea of weeds is to say lightly that 'weed' is merely a relative concept, entirely dependent upon context. This is both true and untrue. Some weeds have such will to power that it is difficult to see them in any other light than as supremacists, a weed in any context. In contrast, weeds have been compared to the concept of the cliché: 'No plants are weeds by nature or by definition. They are weeds if and only if a particular gardener doesn't want them around. One man's uprooted dandelion is another man's dandelion soup.'[1]

And yet . . . to turn this argument back on itself – to double dig, as it were – clichés are clichés because they actually say something true about the world. The most successful clichés, like the most virulent of weeds, are easy to recognize. We may not like them, but we know they exist. The ideal weed is a weed because it has qualities that are weedy in most imagined contexts.

I have touched on only a small proportion of plants that are nearly always considered as weeds, and have investigated in passing their uses, their advantages and their disadvantages. I have conflated weed and wild plant, because the distinction does not hold up. Wild becomes weed, becomes desired plant and again weed, even in one's own garden, even in a single season.

What Alice Sturm says of farming alone is true of all our attempts to interact with plants and weeds in particular. We may constrain and encourage and bend them to our will, 'but it is not an act of

A single small stalwart weed.

creation'.[2] Weeds exist independent of humanity and our attempts to regulate them.

Timeline

c. 3500 BCE	Evidence of common weeds such as bindweed and poppy on Neolithic sites
c. 2800 BCE	Shennong's *Bencao Jing* (Materia Medica) appears, in which combinations of plant extracts dominate
c. 2500–1450 BCE	Images of weeds on frescoes, including nettles, found in the ancient Minoan civilization on the Greek island of Crete
c. 2500–800 BCE	Evidence of weeding and the consumption of weeds in ancient Egypt is found in tombs
370–287 BCE	Theophrastus, Greek naturalist and the Father of Botany, is interested in the allelopathic effects of weeds
c. 300–260 BCE	Theocritus, a Greek poet, mentions 87 different plants including many uncultivated wild ones
23–79 CE	Pliny the Elder, Roman naturalist. His *Historia Naturalis* is an encyclopaedia of knowledge and includes detailed references to plants, including those used for food
50–70 CE	Dioscorides' *De Materia Medica* mentions plants suitable for medicinal use. It becomes a primary text used by the Roman army and an important practical text until the 1500s
c. 70–90	In the Gospel of Matthew appear the parables of the good seed and of the tares

129–216	Life of Galen of Pergamum, Greek physician, interested in what plants peasants used as both herbal remedies and food
c. 140	Apuleius' *Herbarium Apuleii Platonici* appears, produced in Bury St Edmunds Abbey. Parts of the walls remain, harbouring weed plants
1098–1179	St Hildegard, German philosopher, mystic and naturalist, produces two books on natural medicine
1481	*Herbarium Apuleii Platonici* appears, the first printed and illustated edition of the 2nd-century herbal. It is the most widely used remedy book of the Middle Ages
1557	Thomas Tusser produces *500 Points of Good Husbandry*, a calendar of rural practices including weeding
1597	John Gerard, English surgeon and apothecary, publishes a herbal based on that of the Flemish botanist and physician Rembert Dodoens of 1554
1629	John Parkinson, English herbalist, publishes *Paradisi in Sole Paradisus* (A Garden of all Sorts of Pleasant Flowers), describing the cultivation of plants
1640	John Parkinson publishes *Theatrum Botanicum* (The Theatre of Plants), a treatise on plants and their medicinal properties
1653	Nicholas Culpeper, English apothecary and herbalist, publishes *The Complete Herbal*
1753	Carl Linnaeus, Swedish naturalist who applied the Latin binomial naming system for plants, which aims to ensure no two species have the same name, publishes *Species Plantarum*
1790s	The mechanization of agriculture in the West
1821–1897	Life of Sebastian Kneipp, German herbalist and proponent of naturopathic medicine
1857	Charles Darwin's weed patch experiment at Down House, Kent

1859	Publication of Darwin's *On the Origin of Species*
1860s	John Ruskin writes about his ideas of weeds as morally deviant
1914–18	Weed plants on the Western Front grow in the mud of the trenches
1940–41	Weeds thrive in London Blitz bombsites
1945	Johann Künzle, Swiss priest and influential proponent of alternative and herbal medicine, publishes *Das grosse Kräuterheilbuch* (The Great Herbal Medicine Book)
1956	The Weed Science Society of America (WSSA) founded, to encourage and promote the development of knowledge concerning weeds and their impact on the environment
1959	UK Weed Acts introduced, concerning the control of broad-leaved dock, curled dock, common ragwort, creeping thistle and spear thistle, to prevent their spread on private land, with fines of up to £1,000
1965/1973	H. G. Baker's list of the perfect weed's characteristics is published
1970s	Use of chemical herbicides in agriculture begins
1974	The International Weed Science Society formed for the study of weeds and their control, with members from Europe, North America, South America and the Asia-Pacific area
1986	First field trials for genetically engineered crops
1996	Widespread cultivation of glyphosate-resistant crops outside Europe begins
2003	UK Ragwort Control Act introduced

Glossary

aconite, *Aconitum* spp., including *Aconitum napellus*
anemone, wood, lady's nightcap, moonflower, old woman's nest, *Anemone nemorosa*
balsam, Himalayan, policeman's helmet, *Impatiens glandulifera*
barnyard grass, *Echinochloa crus-galli*
betony, bishop's wort, *Stachys officinalis*
bindweed, possession vine, *Convolvulus arvensis*
black-grass, slender meadow foxtail, twitch grass, *Alopecurus myosuroides*
bluebell, common, *Hyacinthoides non-scripta*
bluebell, Spanish, *Hyacinthoides hispanica*
borage, bugloss, *Borago officinalis*
bracken, *Pteridium aquilinum*
bramble, *Rubus fruticosus*
buddleia, butterfly bush, *Buddleia davidii*
burdock, beggar's buttons, fox's clote, thorny burr, love leaves, *Arctium lappa*
buttercup, creeping, *Ranunculus repens*
buttonweed, poorjoe, *Borreria articularis*
camfhur grass, devil weed, Christmas bush, Siam weed, *Chromolaena odorata*
carrot, wild, *Daucus carota*; *see also* cow parsley
castor oil plant, *Ricinus communis*
celandine, lesser, *Ficaria verna*
chickweed, common, starwort, tongue grass, bird weed, *Stellaria media*
Cicely, sweet, *Myrrhis odorata*
cocklebur, *Xanthium strumarium*
cockscomb, *Celosia argentea*
cogon grass, Japanese bloodgrass, *Imperata cylindrica*
coltsfoot, coughwort, *Tussilago farfara*
comfrey, *Symphytum* spp.
cow parsley, Queen Anne's lace, keck, wild chervil, mother-die, wild carrot, *Anthriscus sylvestris*
crabgrass, finger grass, fonio, *Digitaria*
cranesbill, Carolina, Carolina geranium, *Geranium carolinianum*

cuckoo pint, starch root, lords and ladies, *Arum maculatum*
daisy, common, English daisy, bruisewort, bairnwort, *Bellis perennis*
dandelion, swine's snout, prince's crown, *Taraxacum* spp.
darnel, cockle, *Lolium temulentum*
deadly nightshade, *Atropa belladonna*
devil's stinkhorn, devil's stinkpot, *Mutinus elegans*
dock, broad-leaved, *Rumex obtusifolius*
dodder, devil's guts, witch's hair, *Cuscata pentagona*
eelgrass, tape grass, wild celery, *Zostera marina*
eglantine, sweet briar, *Rosa rubiginosa*
elecampane, horse heal, elfwort, *Inula helenium*
enchanter's nightshade, *Circaea lutetiana*
eyebright, *Euphrasia* spp., incl. *Euphrasia officinalis*
fat hen, goosefoot, lamb, *Chenopodium album*
fleabane hairy, rarajeub, *Conyza bonariensis*
fly agaric, *Amanita muscaria*
furze, common gorse, *Ulex europaeus*
gallant soldier, potato weed, *Galinsoga parviflora*
garlic, wild, *Allium ursinum*
giant hogweed, hogsbane, giant cow parsley, *Heracleum mantegazzianum*
goatgrass, *Aegilops geniculata* or *A. ovata*
goosegrass, bedstraw, cleavers, *Galium aparine*
green alkanet, evergreen bugloss, *Pentaglottis sempervirens*
ground elder, herb Gerard, goutweed, bishop's wort, pigweed, *Aegopodium podagraria*
groundsel, old man of the spring, *Senecio vulgaris*
hairy bittercress, *Cardamine hirsuta*
hawkweed, devil's paintbrush, grim-the-collier, *Pilosella aurantiaca*
hawthorn, May blossom, *Crataegus monogyna*
hemlock, break-your-mother's-heart, devil's flower, lady's lace, nosebleed, pickpocket, *Conium maculatum*
henbane, killer of hens, stinking Roger, *Hyoscyamus niger*
herb Robert, jenny wren, pink bird's eye, stinky Bob, *Geranium robertianum*
honeysuckle, wild, woodbine, *Lonicera periclymenum*
horny goatweed, barrenwort, *Epimedium* spp.
horsetail, common or field, mare's tail, *Equisetum arvense*
horseweed, mare's tail, *Conzya canadensis*
hound's tongue, *Cynoglossum officinale*
Job's tears, tear grass, *hata mugi*, Chinese 'pearl barley', *Coix lacryma-jobi*
Johnson grass, *Sorghum halapense*
jungle rice, *Echinochloa colona*
Kahili ginger, *Hedychium gardnerianum*
Kentucky blue grass, *Poa pratensis*
khat, Arabian tea, bushman's tea, *Catha edulis*
knotweed, Japanese, *Fallopia japonica*
kudzu, Japanese, *Pueraria montana*

lady's smock, *Cardamine pratensis*
lamb's quarter, wild spinach, pigweed, goosefoot, *Chenopodium album*
loosestrife purple, *Lythrum salicaria*
lovage, *Levisticum officinale*
lupin, blue, *Lupinus angustifolius*
maple, *Acer* spp.
marigold, marsh, *Caltha palustris*
marigold, Mexican, *Tagetes lucida*
marram grass, beach grass, *Ammophila arenaria*
mayweed, *Anthemis pseudocotula*
meadowsweet, bridewort, *Filipendula ulmaria (Spiraea ulmaria)*
mercury annual, *Mercurialis annua*
mongongo, *Ricinodendron rautanenii*
mustard, *Sinapis alba*
nettle, stinging, hoky-poky, devil's apron, deceiver, naughty man's plaything, *Urtica dioica*
nettle, European, *Urtica pipulifera*
nutsedge, yellow, *Cyperus esculentus*
nutsedge purple, *Cyperus rotundus*
onion, wild, *Allium* spp.
pansy, wild, love-in-idleness, heartsease, *Viola tricolor*
parsley, wild, includes numerous members of the family Apiaceae, incl. *Petroselinum crispum*, garden parsley
Paterson's curse, salvation Jane, *Echium plantagineum*
pimpernel, *Anagallis arvensis*
plantain, *Plantago lanceolata, P. major*
plantain pussytoes, *Antennaria parlinii*
pokeweed, common, *Phytolacca americana*
poppy, opium, *Papaver somniferum*
primrose, Arabian, *Arnebia hispidissima*
primrose, oxlip, *Primula elatior*
purslane, moss rose, pigweed, fat hen, *Portulaca oleracea*
ragwort, Oxford, *Senecio squalidus*
ragwort, stinking Nanny, stinking Willy, St James' wort, *Jacobaea vulgaris*
red clover, *Trifolium pratense*
rhododendron, *Rhododendan ponticum*
rhubarb, Chilean, *Gunnera tinctoria*
rose, wild eglantine, *Rosa rubiginosa*
rose, wild musk, *Rosa moschata*
rue, herb of grace, *Ruta graveolens*
ryegrass, Italian, *Lolium multiflorum*
ryegrass, rigid, *Lolium rigidum*
self-heal, husk head, *Prunella vulgaris*
speedwell, common, figwort, *Veronica persica*
spurge, common or petty, milkweed, cancer weed, *Euphorbia peplus*
spurge, sun, madwoman's milk, *Euphorbia helioscopia*

stitchwort, greater, *Stellaria holostea*
strawberry guava, *Psidium littorale*
tansy, bitter buttons, mugwort, *Tanacetum vulgare*
teasel, *Dipsacus fullorum* and *D. laciniatus*
thale cress, mouse-ear cress, *Arabidopsis thaliana*
toadflax, *Linaria vulgaris*
tormentil, common, furze-bind, *Potentilla erecta*
violet, wild, *Viola odorata*, *V. arvensis*
water betony, green figwort, wasp weed, *Scrophularia umbrosa*
water crowsfoot, *Ranunculus aquatilis*
water hyacinth, *Eichhornia crassipes*
willow, *Salix* spp.
willowherb, great, *Epilobium hirsutum*
willowherb, rosebay, fireweed, great willow-herb, *Epilobium angustifolium*, *Chamerion angustifolium*
withnia nightshade, poison gooseberry, *Withnia somnifera*
wood sorrel, *Oxalis acetosella*
yarrow, field hop, *Achillea millefolium*

References

Introduction

1 *The Taoist Classics*, vol. II: *The Collected Translations of Thomas Cleary* (Boston, MA, 1996), p. 497.
2 Leo Tolstoy, 'The Three Parables', in *The Complete Works of Count Tolstoy* (Bloomington, IN, 1905), vol. XXIII, p. 69.
3 Sabine Durrant, 'I Was Grateful to Her for Dying', *The Guardian* (24 January 2009), 'Family', p. 2.
4 Pranjal Bezbarua et al., 'Management of Invasive Species in Assam, India', 3rd International Symposium on Environmental Weeds and Invasive Plants (Ascona, 2011).
5 Florence Louisa Barclay, *The White Ladies of Worcester: A Romance of the 12th Century* (London, 1917), p. 41.
6 B. R. Trenbath cited in B. D. Booth et al., *Weed Ecology in Natural and Agricultural Systems* (London, 2003), p. 172.
7 Bruce Osborne et al., 'The Riddle of *Gunnera tinctoria* Invasions', 3rd International Symposium on Environmental Weeds and Invasive Plants (Ascona, 2011).
8 William Little and H. C. Fowler, *The New Shorter Oxford Dictionary on Historical Principles* (Oxford, 1973), cited in Robert L. Zimdahl, *Weed Science: A Plea for Thought – Revisited* (London, 2012), p. 7.
9 Zachary J. S. Falck, *Weeds: An Environmental History of Metropolitan America* (Pittsburgh, PA, 2011), p. 175.
10 'Tourists Warned of UAE Laws', www.news.bbc.co.uk, 8 February 2008.
11 R. G. Ellis, cited in *Weed Biology and Management*, ed. S. Inderjit (Dordrecht, 2004), p. 1. A more recent and detailed map of distribution can be found in C. D. Preston et al., *New Atlas of the British and Irish Flora: An Atlas of the Vascular Plants of Britain, Ireland, The Isle of Man and the Channel Islands* (Oxford, 2002).
12 J. A. McNeely, J. O. Luken and J. W. Thieret, cited in *Weed Biology*, ed. Inderjit, p. 2.
13 Rev. Leonard Jenyns, *Memoir of the Rev. John Stevens Henslow, Late Rector of Hitcham and Professor of Botany in the University of Cambridge* [1862] (Cambridge, 2011), pp. 182–3.

14 Nicholas E. Korres, *Encyclopaedic Dictionary of Weed Science: Theory and Digest* (Andover, 2005), pp. 648–9.
15 H. G. Baker, 'The Evolution of Weeds', *Annual Review of Ecology, Evolution and Systematics*, V (1974), pp. 1–24.
16 Inderjit, ed., *Weed Biology*, p. 21.
17 B. D. Booth et al., *Weed Ecology*, p. 241.

1 The Idea of Weeds

1 *The Letters of Horace Walpole, Earl of Orford, including numerous now first published from the original manuscripts, 1778–1797* (London, 1840), vol. VI, p. 57, 10 July 1779.
2 Charles Darwin, *Notebooks on Transmutation of the Species*, ed. Gavin de Beer (London, 1960), October 1838–July 1839 (IV.114). See http://darwin-online.org.uk.
3 Teresa McLean, *Medieval English Gardens* (London, 1981), p. 120.
4 Robert L. Zimdahl, *Weed Science: A Plea for Thought – Revisited* (London, 2012), p. 8.
5 John Ruskin, *Sesame and Lilies* (Rockville, MD, 2008), p. 38.
6 J. C. Loudon cited in Richard Mabey, *Weeds: The Story of Outlaw Plants* (London, 2010), p. 9.
7 J. C. Chacón and S. R. Gliessman, 'Use of "Non-Weed" Concept in Traditional Tropical Agroecosystems of South-Eastern Mexico', in *Agro-Ecosystems*, VIII (1982), pp. 1–11.
8 D. E. De Pietri cited in Diane Sage et al., 'Effect of Grazing Exclusion on the Woody Weed *Rosa rubiginosa* in High Country Short Tussock Grasslands', *New Zealand Journal of Agricultural Research*, LII (2009), p. 126.
9 Michael Pollan, *A Plant's Eye View*, TED talk, www.ted.com, 20 August 2012.
10 John Roach, '2,000-Year-Old Seed Sprouts, Sapling is Thriving', *National Geographic News*, http://news.nationalgeographic.com, 22 November 2005.
11 Michael Pollan, 'Gardening Means War', *New York Times Magazine* (19 June 1988).
12 W. Hilbig cited in *Weed Biology and Management*, ed. S. Inderjit (Dordrecht, 2004).

2 The Background

1 Nigel F. Hepper, *Pharaoh's Flowers: The Botanical Treasures of Tutankhamun* (London, 1990), p. 18.
2 J. E. Raven and Faith Raven, *Plants and Plant Lore in Ancient Greece* (Oxford, 2000), p. 26.
3 Al-Farabi, cited in Michael S. Kochin, 'Weeds: Cultivating the Imagination in Medieval Arabic Political Theology', *Journal of the History of Ideas* (1999), pp. 399–416.
4 Ibn Bajjah, *Governance of the Solitary*, in Kochin, 'Weeds', p. 402.

5 Sue Shephard, *The Surprising Life of Constance Spry* (London, 2010), cited in Joanna Fortnam, 'Society Florist Constance Spry remembered in Mayfair', *The Telegraph* (13 November 2011).
6 Oliver Leaman, email to author, 20 May 2014.
7 Quoted in Ralph Waldo Emerson, 'Thoreau', *The Atlantic* (August 1862). See www.theatlantic.com.
8 Henry David Thoreau cited in Michael Pollan, 'Weeds Are Us', *New York Times* (5 November 1989).
9 Ibid.
10 Charles Darwin, *Notebooks on Transmutation of the Species*, ed. Gavin de Beer (London, 1960), October 1838–July 1839 (IV.114). See http://darwin-online.org.uk.
11 Letter from Darwin to Joseph Hooker, 12 April 1857, in *Darwin's Letters: Collecting Evidence*, www.pbs.org.
12 Ibid., p. 179.
13 Mario Livio, *Brilliant Blunders: From Darwin to Einstein – Colossal Mistakes by Great Scientists that Changed our Understanding of Life and the Universe* (London, 2014), cited in Freeman Dyson, 'The Case for Blunders', in *New York Review of Books* (6 March 2014).
14 Gregor Mendel, 'Experiments in Plant Hybridization' (originally *Versuche über Pflanzen-Hybriden*), *Journal of the Brünn Natural History Society* (1865), see http://esp.org.
15 Japanese Knotweed Control, www.japaneseknotweedcontrol.com.
16 Jonathan Gressel, *Crop Ferality and Volunteerism* (London, 2005), p. 3.
17 Richard Mabey, *Weeds: The Story of Outlaw Plants* (London, 2010), p. 23.
18 John Gerard, cited ibid., p. 91.
19 John Hersey, *Hiroshima* [1946] (London, 2001), p. 91.
20 Dorothy Sterling et al., *The Outer Lands* (New York, 1992), pp. 63–4.

3 Image and Allegory

1 Stella Gibbons, *Cold Comfort Farm* [1932] (London, 2006), p. 32.
2 Charles Seddon Evans, *The Sleeping Beauty* (London, n.d.).
3 St Augustine, Sermon 23 on the New Testament. See www.newadvent.org/fathers/160323.htm.
4 Martin Luther, *Martin Luther's Postil* (Macomb, MI, 2012), vol. I, p. 216.
5 John Milton, *Areopagitica: A Speech for the Liberty of Unlicensed Printing to the Parliament of England* (London, 1644). See www.gutenberg.org.
6 Martin Luther's *Tischreden*, *Table Talk*, c. 1532, cited in Julia Hughes-Jones, *The Secret History of Weeds: What Women Need to Know About Their History* (Bradenton, FL, 2009).
7 M. M. Mahood, *The Poet as Botanist* (Cambridge, 2008), p. 11.
8 Judith Bronkhurst, *William Holman Hunt: A Catalogue Raisonné* (London, 2006), vol. I, p. 153.
9 H. W. Holman, *Pre-Raphaelitism and the Pre-Raphaelite Brotherhood* (London, n.d.), p. 350.

10 Bronkhurst, *William Holman Hunt*, p. 158.
11 Hans Christian Adam, *Karl Blossfeldt, 1865–1932* (Cologne, 2001), p. 26.
12 Richard Mabey, 'The Lowly Weed Has Its Day', *Tate Etc.*, 22 (Summer 2011).
13 Ibid.
14 David Blayney Brown, 'Draughtsman and Watercolourist', in *J.M.W. Turner: Sketchbooks, Drawings and Watercolours*, ed. Brown, www.tate.org.uk, December 2012.
15 John Ruskin and Clive Wilmer, *Unto This Last and Other Writings* (London, 1985), p. 172.
16 Kathryn Shattuck, 'Leaves Speak; A Journalist Listens', *New York Times* (20 July 2008).
17 Patrick Dougherty, personal email, March 2014.
18 Hugo Worthy, 'Jacques Nimki: I Want Nature', *Interface*, www.a-n.co.uk, accessed 11 December 2013.
19 Jemima Montagu, 'The Art of the Garden', *Tate Etc.*, 1 (Summer 2004).
20 Cited in Louisa Buck, 'Champion of the Urban Weed', *The Art Newspaper* (December 2002).
21 Ovid, cited in 'Aconite Poisoning', http://penelope.uchicago.edu.
22 August Strindberg, *Miss Julie* [1888], trans. Michael Meyer (London, 1964), 'Pantomime'.
23 Henry David Thoreau, 'The Bean Field', in *Walden: An Annotated Edition*, ed. Walter Harding (New York, 1995).
24 Mina Gorji, 'John Clare's Weeds', in *Ecology and the Literature of the British Left: The Red and the Green*, ed. John Rignall and H. Gustav Klaus (Farnham, 2012), p. 73.
25 Zachary Falck, 'Weeds: An Environmental History of Metropolitan America', *American Historical Review*, CXVII/5 (2012), p. 1619.
26 Hans Christian Andersen, 'The Wild Swans' (1838), in *The Harvard Classics*, www.bartleby.com.
27 Herman Melville, *Billy Budd, Sailor and Selected Tales* (Oxford, 2009), p. 362.

4 Unnatural Selection: The War on Weeds

1 The Jatan Trust, Gujarat, India, an organic farming movement set up to counter the use of chemicals in farming.
2 Jonathan Bate, *John Clare: A Biography* (London, 2003), p. 23.
3 Virgil, *The Georgics*, I, trans. A. S. Kline (2002), 'The Beginnings of Agriculture', pp. 118–59. See www.poetryin translation.com.
4 Alice Sturm, 'Weeds', *The Hypocrite Reader*, 17, 'Hide and Seek' (June 2012), www.hypocritereader.com.
5 Pehr Kalm, *Pehr Kalm's Visit to England, On His Way to America in 1748*, trans. Joseph Lucas (London, 1892), p. 174.
6 Ibid., p. 353.
7 Cited in Lorne Clinton Evans, *Weeds in the Prairie West: An Environmental History* (Calgary, 2002), p. 30.

8 Jon Tourney, 'Grapegrowers Face Herbicide-resistant Weeds', www.winesandvines.com, 20 June 2011.
9 M. R. Sabbatini et al., 'Vegetation: Environmental Relationships in Irrigation Channel Systems of Southern Argentina', *Aquatic Botany*, 60 (1998), pp. 119–33, and O. A. Fernández et al., 'Interrelationships of Fish and Channel Environmental Conditions with Acquatic Macrophytes in an Argentine Irrigation System', *Hydrobiologia*, 380 (1998), pp. 15–25.
10 S. Inderjit, ed., *Weed Biology and Management* (Dordrecht, 2004), p. 117.
11 Pat Michalak, 'Use Tadpole Shrimps to Control Weeds in Transplanted Paddy Rice', Rodale Institute, www.newfarm.org, accessed 1 December 2013.
12 Amy Stewart, *Wicked Plants: The Weed that Killed Lincoln's Mother and Other Botanical Atrocities* (Chapel Hill, NC, 2009), p. 148.
13 Ibid., p. 75.
14 Liz Taylor on BBC Radio 4, *Thinking Allowed* (15 April 2014), 'British Working Class Gardens'.
15 Cited in Richard Mabey, *Weeds: The Story of Outlaw Plants* (London, 2010), pp. 65–6.
16 Cited in Evans, *Weeds in the Prairie West*, p. 28.
17 Sturm, 'Weeds'.
18 Michael Le Page, 'Unnatural Selection: Wild Weeds Outwit Herbicides', *New Scientist* (May 2011).
19 Daniel Chamovitz, *What a Plant Knows: A Field Guide to the Senses of Your Garden and Beyond* (London, 2013), p. 43.
20 Jed B. Colquhoun, 'Allelopathy in Weeds and Crops: Myths and Facts' (2006), www.soils.wisc.edu.
21 Peter J. Bowden in *Agrarian History of England and Wales*, ed. E.J.T. Collins (Cambridge, 2000), cited in Evans, *Weeds in the Prairie West*, p. 29.
22 Matt Jenkins, 'Pacific Invasion', *Nature Conservancy* (October 2013), pp. 49–59.
23 Ibid. p. 58.
24 John Payne, 'Reviewing *Farmerbots: A New Industrial Revolution* by James Mitchell Crow', www.robohub.org, 11 November 2012.
25 Svend Christensen of the Danish Institute of Agricultural Sciences at Tjele, cited in Duncan Graham-Rowe, 'Weedkilling Robots Slash Herbicide Use', *New Scientist* (June 2003).
26 'Precision Herbicide Drones Launch Strikes on Weeds', *New Scientist* (July 2013).
27 Douglas Buhler of the United States Department of Agriculture, cited in Jonathan Beard, 'How to Let Sleeping Weed Seeds Lie', *New Scientist* (June 1995).
28 Andy Coghlan, 'Weeds Get Boost from GM Crops', *New Scientist* (August 2002).
29 Andy Coghlan, 'Master Gene Helps Weeds Defy All Weedkillers', *New Scientist* (March 2013).

30 Stephanie Pain, 'A Giant Leap for Plantkind', in 'Cuttings: A Round-up of the Latest Plant Science Stories', *Kew Magazine* (Spring 2014).

5 Useful Weeds

1 Cited in Kristina Lerman, 'The Life and Works of Hildegard von Bingen, 1098–1179', www.fordham.edu, 15 February 1995.
2 See www.botanical.com., accessed 10 December 2013.
3 'Camden' cited in Lady Wilkinson, *Weeds and Wild Flowers: Their Uses, Legends and Literature* (London, 1858), p. 8.
4 Aaron Hill, *Aaron Hill's Works* (London, 1753), vol. IV, p. 92.
5 John Evelyn, *Acetaria: A Discourse of Sallets* (1699), http://gutenberg.org/ebooks, pp. 88–9.
6 Ole Peter Grell, 'Medicine and Religion in Sixteenth Century Europe', in *The Healing Arts: Health, Disease and Society in Europe, 1500–1800*, ed. Peter Elmer (Manchester, 2014), p. 90.
7 Peter Elmer, 'The Care and Cure of Mental Illness', in *The Healing Arts*, ed. Elmer, p. 239.
8 Matthaeus Sylvaticus, a physician of Mantua, in George Don, *A General History of the Dichleamydeous Plants . . . Arranged According to the Natural System* (London, 1838), vol. IV, p. 610.
9 Lady Mildmay cited in Jennifer Wynne Hellwarth, 'Be Unto Me as a Precious Ointment: Lady Grace Mildmay, Sixteenth-century Female Practitioner', *Acta Hisp. Med. Sci. Hist. Rlus*, 19 (1999), p. 115. See http://ddd.uab.cat.
10 'An Approved Conserve for a Cough or Consumption of the Lungs', in Anon., *A Queen's Delight or the Art of Preserving, Conserving and Candying, as also A Right Knowledge of Making Perfumes, and Distilling the Most Excellent Waters* [1671]. See http://gutenberg.org/ebooks/15019.
11 'Conserves of Violets, the Italian Manner', ibid.
12 'South African Geranium Root may Kill HIV-1', www.financialexpress.com, 31 January 2014.
13 J. E. Raven and Faith Raven, *Plants and Plant Lore in Ancient Greece* (Oxford, 2000), p. 34.
14 John Gerard, cited in D. C. Watts, *Dictionary of Plant Lore* (Philadelphia, PA, 2007), p. 171.
15 John Gerard, cited in 'Goutweed', www.botanical.com., accessed 13 January 2013.
16 *Lacnunga*, a collection of Anglo-Saxon medical texts and prayers, no. 82 in Eleanour Sinclair Rohde, *The Old English Herbals* [1922], http://gutenberg.org.
17 John Gerard, cited in Pamela Jones, *Just Weeds: History, Myths and Uses* (Boston, MA, 1994), p. 20.
18 Cited in Stewart, *Wicked Plants*, p. 32.
19 Cited in Jonathan Bate, *John Clare: A Biography* (London, 2003), p. 97.
20 Cited in Lady Wilkinson, *Weeds and Wild Flowers*, p. 5.

21 H. D. Nelson et al., 'Nonhormonal Therapies for Menopausal Hot Flashes: Systematic Review and Meta-analysis', *Journal of the American Medical Association* (2006), pp. 2057–71.

6 In Our Diet

1 Richard Mabey, *Food for Free* (London, 2012), p. 8.
2 Richard Lee, *The !Kung San: Men, Women and Work in a Foraging Society* (Cambridge, 1979), p. 94.
3 'The Bushmen Call it Mongongo', www.elephantswithoutborders.org, 19 November 2010.
4 Wendell Berry, 'The Pleasures of Eating', in *The Art of the Commonplace: The Agrarian Essays of Wendell Berry*, ed. Norman Wirzba (Berkeley, CA, 2004), p. 321.
5 'Wild Garlic, Nettle and Bittercress Pesto', *Robin Harford's Wild Food Guide to the Edible Plants of Britain*, www.eatweeds.co.uk, accessed 4 April 2014.
6 'Lesser Celandine and Ground Ivy Stew', ibid.
7 Paul Peacock, *The Pocket Guide to Wild Food* (Preston, 2008), p. 34.
8 Roslynn Brain and Hayley Waldbillig, 'Urban Edibles: Weeds', *Utah State University Extension Sustainability*, www.extension.usu.edu, February 2013.
9 Oliver Strand, 'A New Leaf', American *Vogue* (May 2013).
10 Ibid.
11 *The Guardian* (24 June 2011).
12 See 'Ground Elder Quiche', at www.eatweeds.co.uk, accessed 6 October 2014.
13 Gail Harland, *The Weeder's Digest: Identifying and Enjoying Edible Weeds* (Totnes, 2012), p. 150.
14 Adapted from Brain and Waldbillig, 'Urban Edibles: Weeds'.
15 John Evelyn, *Acetaria: A Discourse of Sallets* (1699), http://gutenberg.org/ebooks.
16 Cited in Pamela Jones, *Just Weeds: History, Myths and Uses* (Boston, MA, 1994), p. 213.
17 Harland, *The Weeder's Digest*, pp. 129 and 151.
18 Adapted from Tristan Stephenson, 'Dandelion and Burdock', The Good Food Channel, http://uktv.co.uk, accessed 14 November 2014.
19 Michael Pollan, *The Omnivore's Dilemma* (New York, 2007), p. 280.
20 'Food Foraging in South Africa', www.foodandthefabulous.com, February 2014.
21 Ibid.

7 A Wild and Weedy Garden

1 Michael Pollan, 'Weeds Are Us', *New York Times* (5 November 1989).
2 William Robinson, *The Garden: An Illustrated Weekly Journal of Gardening in all its Branches*, II (1872).

3 William Robinson, *The Wild Garden: Or, Our Groves and Shrubberies Made Beautiful* [1870] (Cork, 2010), p. 76.

4 Rev. Leonard Jenyns, *Memoir of the Rev. John Stevens Henslow, Late Rector of Hitcham and Professor of Botany in the University of Cambridge* [1862] (Cambridge, 2011), p. 189.

5 Gertrude Jekyll, *Home and Garden: Notes and Thoughts, Practical and Critical, of a Worker in Both* [1890] (Cambridge, 2011), p. 277.

6 Oliver Strand, 'A New Leaf', American *Vogue* (May 2013).

7 See www.thehighline.org, accessed 6 October 2014.

8 Edward Draper, 'Do Motorways Create a Microclimate?', www.nakedscientists.com, 6 June 2012.

9 David Quammen, 'Planet of Weeds', *Harper's* (October 1998).

10 David Quammen, *Natural Acts: A Sidelong View of Science and Nature* (New York, 2009), p. 188.

Afterword

1 Allan Metcalf, 'Garden-variety Clichés', *Chronicle of Higher Education Review* (14 March 2014).

2 Alice Sturn, 'Weeds', *The Hypocrite Reader*, 17 (June 2012).

Further Reading

Adam, Hans Christian, *Karl Blossfeldt, 1865–1932* (Cologne, 2001)

Anderson, Rohan, *Whole Larder Love* (New York, 2012)

Berry, Wendell, *The Art of the Commonplace: The Agrarian Essays of Wendell Berry* (Berkeley, CA, 2004)

Booth, B. D., S. D. Murphy and C. J. Swanton, *Weed Ecology in Natural and Agricultural Systems* (London, 2003)

Bronkhurst, Judith, *William Holman Hunt: A Catalogue Raisonné* (London, 2006)

Chamovitz, Daniel, *What a Plant Knows: A Field Guide to the Senses of Your Garden and Beyond* (London, 2013)

Chatto, Beth, *Woodland Gardens* (London, 2002)

Colquhoun, Jed B., 'Allelopathy in Weeds and Crops: Myths and Facts', www.soils.wisc.edu (2006)

De Bray, Lys, *The Wild Garden: An Illustrated Guide to Weeds* (London, 1978)

Edmonds, William, *Weeds, Weeding (& Darwin): The Gardeners' Guide* (London, 2013)

Elmer, Peter, ed., *The Healing Arts: Health, Disease and Society in Europe, 1500–1800* (Manchester, 2014)

Evans, Clinton Lorne, *Weeds in the Prairie West: An Environmental History* (Calgary, 2002)

Flowerdew, Bob, *Go Organic!* (London, 2002)

Gibbons, Euell, *Stalking the Wild Asparagus* (Chambersburg, PA, 1962)

Gorji, Mina, 'John Clare's Weeds', in John Rignall and H. Gustav Klaus, eds, *Ecology and the Literature of the British Left: The Red and the Green* (Farnham, 2012)

Gressel, Jonathan, ed., *Crop Ferality and Volunteerism* (London, 2005)

Haragan, Patricia Dalton, *Weeds of Kentucky and Adjacent States* (Lexington, KY, 1953)

Harland, Gail, *The Weeder's Digest: Identifying and Enjoying Edible Weeds* (Totnes, 2012)

Hepper, F. Nigel, *Pharaoh's Flowers: The Botanical Treasures of Tutankhamun* (London, 1990)

Holladay, Harriett McDonald, *Kentucky Wildflowers* (Lexington, KY, 1956)

Inderjit, S., ed., *Weed Biology and Management* (Dordrecht, 2004)

Jekyll, Gertrude, *Home and Garden: Notes and Thoughts, Practical and Critical, of a Worker in Both* [1890] (Cambridge, 2011)

Jenyns, Rev. Leonard, *Memoir of the Rev. John Stevens Henslow, Late Rector of Hitcham and Professor of Botany in the University of Cambridge* [1862] (Cambridge, 2011)

Jones, Pamela, *Just Weeds: History, Myths and Uses* (Boston, MA, 1994)

Kallas, John, *Edible Wild Plants: Wild Foods from Dirt to Plate* (Layton, UT, 2010)

Kalm, Pehr, *Pehr Kalm's Visit to England, On His Way to America in 1748*, trans. Joseph Lucas (London, 1892)

Kochin, Michael S., 'Weeds: Cultivating the Imagination in Medieval Arabic Political Theology', *Journal of the History of Ideas*, LX/3 (July 1999), pp. 399–416

Korres, Nicholas E., *Encyclopaedic Dictionary of Weed Science: Theory and Digest* (Andover, 2005)

Laws, Bill, *Fifty Plants that Changed the Course of History* (New York, 2011)

Lee, Richard, *The !Kung San: Men, Women and Work in a Foraging Society* (Cambridge, 1979)

Livio, Mario, *Brilliant Blunders: From Darwin to Einstein – Colossal Mistakes by Great Scientists that Changed our Understanding of Life and the Universe* (London, 2014)

Mabey, Richard, *Food for Free* [1972] (London, 2012)

—, *Weeds: The Story of Outlaw Plants* (London, 2010)

McClean, Teresa, *Medieval English Gardens* (London, 1981)

Mahood, M. M., *The Poet as Botanist* (Cambridge, 2008)

Negbi, Moshe, 'A Sweetmeat Plant, a Perfume Plant and their Weedy Relatives: A Chapter in the History of *Cyperus esculentus L.* and *C. rotundus L.*', *Economic Botany*, XLVI/1 (1992), pp. 64–71

Newton, John, *The Roots of Civilisation: Plants that Changed the World* (London, 2009)

Peacock, Paul, *The Pocket Guide to Wild Food* (Preston, 2008)

Pfeiffer, Ehrenfried, *Weeds and What They Tell Us* (London, 2012)

Phillips, Roger et al., *Garden and Field Weeds* (London, 1986)

Pochin, Eric, *A Nature Lover's Note Book* (Leicester, 1944)

Pollan, Michael, *The Omnivore's Dilemma* (New York, 2007)

Preston, C. D., D. A. Pearman and T. D. Dines, *New Atlas of the British and Irish Flora: An Atlas of the Vascular Plants of Britain, Ireland, The Isle of Man and the Channel Islands* (Oxford, 2002)

Raven, J. E., and Faith Raven, *Plants and Plant Lore in Ancient Greece* (Oxford, 2000)

Readman, Jo, *Weeds: How to Control and Love Them* (London, 1991)

Robinson, William, *The Wild Garden: Or, Our Groves and Shrubberies Made Beautiful* [1870] (Cork, 2010)

Rose, Francis, *The Observer's Book of Wild Flowers* (London, 1983)

Rose, Graham, *The Traditional Garden Book* (London, 1993)

Ruskin, John, and Clive Wilmer, *Unto This Last and Other Writings* (London, 1985)

Shannon, Nomi, *The Raw Gourmet* (Burnaby, BC, 2007)

Spencer, Edwin Rollin, *All About Weeds* (New York, 2011)

Stein, Sara B., *My Weeds: A Gardener's Botany* (Gainsville, FL, 2000)
Sterling, Dorothy, Robert Finch and Winifred Lubell, *The Outer Lands* (New York, 1992)
Stewart, Amy, *Wicked Plants: The Weed that Killed Lincoln's Mother and Other Botanical Atrocities* (Chapel Hill, NC, 2009)
Sturm, Alice, 'Weeds', *Hypocrite Reader*, 17 (June 2012)
Thayer, Samuel, *The Forager's Harvest: A Guide to Identifying, Harvesting and Preparing Edible Wild Plants* (Ogema, WI, 2006)
Walker, Barbara M., *The Little House Cookbook: Frontier Foods from Laura Ingalls Wilder's Classic Stories* (New York, 1989)
Watson, Bob, *Plants: Their Use, Management, Cultivation and Biology* (Ramsbury, Wiltshire, 2008)
Weber, E., *Invasive Plant Species of the World: A Reference Guide to Environmental Weeds* (London, 2003)
Whitfield, Roderick, *Fascination of Nature: Plants and Insects in Chinese Paintings and Ceramics of the Yuan Dynasty (1279–1368)* (London, 1986)
Wilkinson, Lady, *Weeds and Wild Flowers: Their Uses, Legends and Literature* (London, 1858)
Willes, Margaret, *The Gardens of the British Working Class* (New Haven, CT, 2014)
Zimdahl, Robert L., *Weed-crop Competition: A Review* (Oxford, 2004)
—, *Weed Science: A Plea for Thought – Revisited* (London, 2012)

Associations and Websites

A MODERN HERBAL
The medicinal, culinary, cosmetic and economic properties of plants, their cultivation and folklore. First published in 1931 by Mrs M. Grieve
www.botanical.com

EAT WEEDS
Wild food guide to the edible plants of the UK, by Robin Harford
www.eatweeds.co.uk

GARDEN ORGANIC
UK organic growing charity, dedicated to researching and promoting organic gardening, farming and food
www.gardenorganic.org.uk

GARDEN WITHOUT DOORS
Contains a UK weed identification guide for garden weeds
www.gardenwithoutdoors.org.uk

GERTRUDE JEKYLL GARDEN
Gertrude Jekyll's gardens at Upton Grey, Hampshire, including the wild garden
www.gertrudejekyllgarden.co.uk

GUERRILLA GARDENING
The guerrilla garden at Elephant and Castle, London – and promoting other city schemes
www.guerrillagardening.org

THEOI
Guides to the plants and flowers of Greek mythology
www.theoi.com/Flora1.html; www.theoi.com/Flora2.html

WILDFLOWER FINDER
Wildflower identification guide
www.wildflowerfinder.org.uk

Acknowledgements

My thanks to Kaye Aboitiz, Dorothy Bank, Judith Bronkhurst, Brigitte Dold, Patrick Dougherty, Peter Edwards, Josie Floyd, Anne Griffin of Kew Gardens Library, Kelly Landen of Elephants Without Borders, Michael Landy, the National Art Library, Oliver Leaman, Lesley Manning, Jayj Moi, Jacques Nimki, Eli Resvanis, Richard Reynolds of guerrillagardening.org, Rosamund Wallinger, the Asia department of the British Museum, the British Library, the Wellcome Library, the Royal Botanic Gardens, Kew Library, the Lindley Library, London, the Royal Horticultural Society, the Victoria & Albert Museum, the Garden Museum, my editor Martha Jay and Michael Leaman.

Photo Acknowledgements

The author and the publishers wish to express their thanks to the below sources of illustrative material and/or permission to reproduce it.

© The Trustees of the British Museum, London: pp. 10, 36, 44, 69, 74 bottom; Patrick Dougherty: pp. 82–3, 84; Nina Edwards: pp. 51, 56, 60, 121, 124, 174, 187, 190–91, 195, 199, 202, 206; The J. Paul Getty Museum / The Getty, Los Angeles: pp. 20–21, 49, 54, 72, 111, 113, 194, 196–7; Michael Landy: p. 90; Library of Congress, Washington, DC: pp. 29, 59, 116–17, 120, 127, 176, 200; Lesley Manning: pp. 6, 13, 14, 27, 41, 70, 98, 100–101, 106–7, 109, 129, 132–3, 139, 153, 186; National Gallery of Art, Washington, DC: pp. 9, 30 bottom, 42, 192, 198; National Library of Medicine, Betheseda: pp. 19, 114, 123, 141, 146, 149, 150, 155, 157, 158; Jacques Nimki: pp. 87, 88–9, 173; Royal Collection Trust, Windsor: p. 74 top; Danuta Solowiej: p. 79; Victoria & Albert Museum, London: pp. 15, 23, 28, 30 top, 34, 43, 46, 47, 48, 55, 65, 76, 93, 95, 115, 142, 175, 204; Wellcome Images, London: pp. 16, 33, 38, 50, 62, 92, 99, 108, 138, 140, 144, 145, 147, 152, 154, 159, 160, 162, 165, 166–7, 170–71, 179, 180, 182, 201.

Index